AF560372

VECTOR ANALYSIS AND COORDINATE GEOMETRY

VECTOR ANALYSIS AND COORDINATE GEOMETRY

Dr. Amit Kumar

RANDOM PUBLICATIONS

NEW DELHI - 110 002 (INDIA)

Vector Analysis and Coordinate Geometry

ISBN 978-93-51117-12-4

Published in 2015 in India by

RANDOM PUBLICATIONS

4376-A/4B, Gali Murari Lal, Ansari Road
New Delhi-110 002
Phone: +9111-43580356, 23289044
E-mail: randomexports@gmail.com; sales@randompublications.com; info@randompublications.com

Reprint 2022

Type Setting by: Friends Media, Delhi-110089
Digitally Printed at: Replika Press Pvt. Ltd.

Preface

A vector is an object that has both a magnitude and a direction. Geometrically, we can picture a vector as a directed line segment, whose length is the magnitude of the vector and with an arrow indicating the direction. The direction of the vector is from its tail to its head. A vector is a specific mathematical structure. It has numerous physical and geometric applications, which result mainly from its ability to represent magnitude and direction simultaneously. Wind, for example, had both a speed and a direction and, hence, is conveniently expressed as a vector. The same can be said of moving objects and forces.

Vector analysis is a branch of mathematics that deals with quantities that have both magnitude and direction. The basic objects in vector calculus are scalar fields (scalar-valued functions) and vector fields (vector-valued functions). Recall that a plane is a flat surface that goes on forever in both directions. If we were to place a point on the plane, coordinate geometry gives us a way to describe exactly where it is by using two numbers. The term "vector analysis" is sometimes used as a synonym for the broader subject of multivariable calculus, which includes vector calculus as well as partial differentiation and multiple integration. Vector calculus plays an important role in differential geometry and in the study of partial differential equations. It is used extensively in physics and engineering, especially in the description of electromagnetic fields, gravitational fields and fluid flow. Vector calculus was developed from quaternion analysis by J. Willard Gibbs and Oliver Heaviside near the end of the 19th century, and most of the notation and terminology was established by Gibbs and Edwin Bidwell Wilson in their 1901 book, *Vector Analysis*. In the conventional form using cross products, vector calculus does not generalize to higher dimensions, while the alternative approach of geometric algebra, which uses exterior products does generalize. A coordinate system is a system which uses one or more numbers, or coordinates, to uniquely determine the position of a point or other geometric element on a manifold such as Euclidean space. The term coordination geometry is used in a number of related fields of chemistry and solid state chemistry/physics. In algebra, we were introduced to the coordinate system, plotting ordered pairs, and graphing lines. These tools are used in geometry as well. Algebra and geometry are used hand-in-hand to solve many real-world math problems. Points, lines, line segments, rays, and angles, as well as other geometric shapes, can be graphed on the coordinate plane. In the field of inorganic coordination complexes it is the geometrical pattern formed by the atoms in the ligands that are bonded to the central atom in a molecule or a coordination complex. The geometrical arrangement will vary according to the number and type of ligands bonded to the metal centre, and to the coordination preference of the central atom, typically a metal in a coordination complex. The number of atoms bonded, (i.e. the number of σ-bonds between central atom and ligands) is termed the coordination number. The geometrical pattern can be described as a polyhedron where the vertices of the polyhedron are the centres of the coordinating atoms in the ligands. The coordination preference of a metal often varies with its oxidation state. One of the most common coordination geometries is octahedral, where six ligands are coordinated to the metal in a symmetrical distribution, leading to the formation of an octahedron if lines were drawn between the ligands. Other common coordination geometries are tetrahedral and square planar.

When employed with skill and understanding, vector analysis can be a practical and powerful tool. Though, written to meet the requirements of graduate students, it will also be useful to physicists and engineers.

I thank all members of my team who have helped in the preparation of the book. My special thanks go to "Random Publications" who have published the book.

— Dr. Amit Kumar

Contents

Chapter 1

Introduction

Since empirical laws of physics do not depend on special or accidental choices of reference frames selected to represent physical relations and geometric configurations, vector analysis forms an ideal tool for the study of the physical universe. The introduction of a special reference frame or coordinate system establishes a correspondence between vectors and sets of numbers representing the components of vectors in that frame, and it induces definite rules of operation on these sets of numbers that follow from the rules for operations on the line segments.

Vector Algebra

A vector is an object that has both a magnitude and a direction. Geometrically, we can picture a vector as a directed line segment, whose length is the magnitude of the vector and with an arrow indicating the direction. The direction of the vector is from its tail to its head.

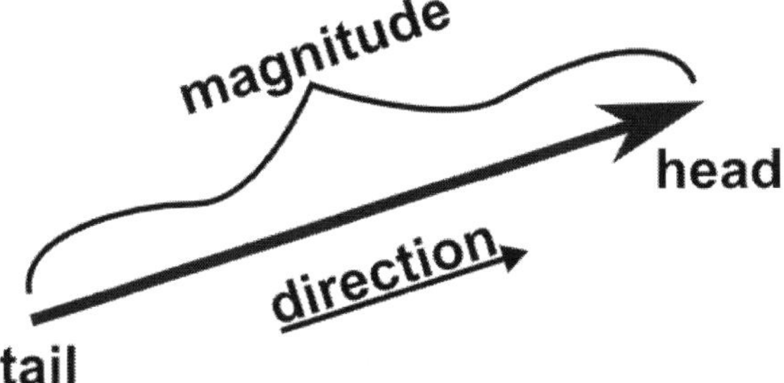

Two vectors are the same if they have the same magnitude and direction. This means that if we take a vector and translate it to a new position (without rotating it), then the vector we obtain at the end of this process is the same vector we had in the beginning.

Two examples of vectors are those that represent force and velocity. Both force and velocity are in a particular direction. The magnitude of the vector would indicate the strength of the force or the speed associated with the velocity.

We denote vectors using boldface as in a or b. Especially when writing by hand where one cannot easily write in boldface, people will sometimes denote vectors using arrows as in $a\rightarrow$ or $b\rightarrow$, or they use other markings. We won't need to use arrows here. We denote the magnitude of the vector a by $\| \mathbf{a} \|$. When we want to refer to a number and stress that it is not a vector, we can call the number a scalar. We will denote scalars with italics, as in a or b.

You can explore the concept of the magnitude and direction of a vector using the below applet. Note that moving the vector around doesn't change the vector, as the position of the vector doesn't affect the magnitude or the direction. But if you stretch or turn the vector by moving just its head or its tail, the magnitude or direction will change.

There is one important exception to vectors having a direction. The zero vector, denoted by a boldface 0, is the vector of zero length. Since it has no length, it is not pointing in any particular direction. There is only one vector of zero length, so we can speak of the zero vector.

Operations on Vectors

We can define a number of operations on vectors geometrically without reference to any coordinate system. Here we define addition, subtraction, and multiplication by a scalar. On separate pages, we discuss two different ways to multiply two vectors together: the dot product and the cross product.

Addition of Vectors

Given two vectors a and b, we form their sum a+b, as follows. We translate the vector b until its tail coincides with the head of a. (Recall such translation does not change a vector.) Then, the directed line segment from the tail of a to the head of b is the vector a+b.

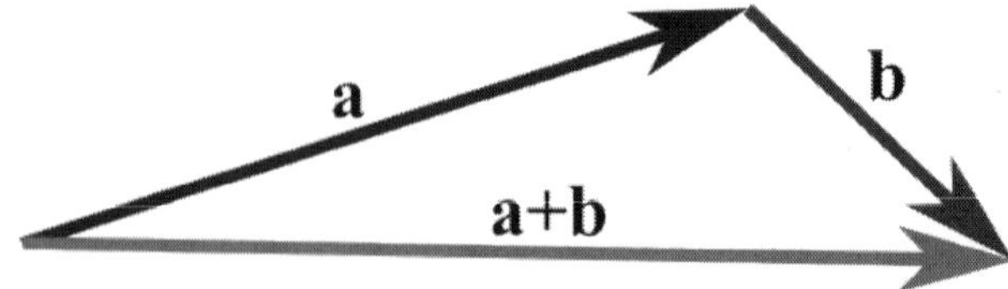

The vector addition is the way forces and velocities combine. For example, if a car is travelling due north at 20 miles per hour and a child in the back seat behind the driver throws an object at 20 miles per hour toward his sibling who is sitting due east of him, then the velocity of the object (relative to the ground!) will be in a north-easterly direction. The velocity vectors form a right triangle, where the total velocity is the hypotenuse.

Therefore, the total speed of the object (i.e., the magnitude of the velocity vector) is 202+202--------√=202√ miles per hour relative to the ground.

Addition of vectors satisfies two important properties.

1. The commutative law, which states the order of addition doesn't matter:

   a+b=b+a.

   This law is also called the parallelogram law, as illustrated in the below image. Two of the edges of the parallelogram define a+b, and the other pair of edges define b+a. But, both sums are equal to the same diagonal of the parallelogram.

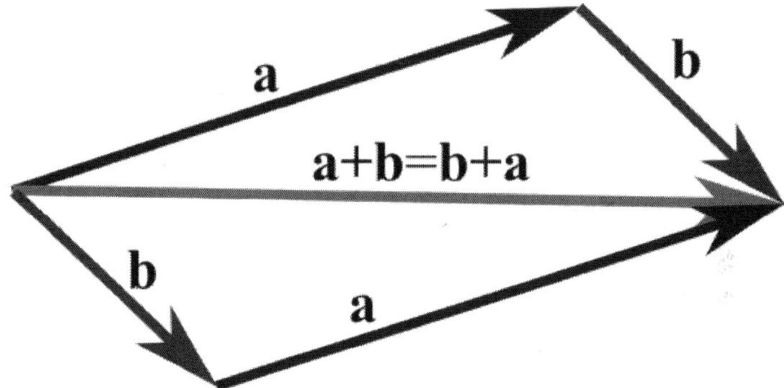

2. The associative law, which states that the sum of three vectors does not depend on which pair of vectors is added first:

   (a+b)+c=a+(b+c).

   You can explore the properties of vector addition with the following applet.

The sum of two vectors. The sum a+b of the vector a and the vector b is shown by the green arrow. As vectors are independent of their starting position, both blue arrows represent the same vector a and both red arrows represent the same vector b. The sum a+b can be formed by placing the tail of the vector b at the head of the vector a. Equivalently, it can be formed by placing the tail of the vector a at the head of the vector b. Both constructions together form a parallelogram, with the sum a+b being a diagonal. (For this reason, the commutative law a+b=b+a is sometimes called the parallelogram law.)

Vector Subtraction

Before we define subtraction, we define the vector “a, which is the opposite of a. The vector “a is the vector with the same magnitude as a but that is pointed in the opposite direction.

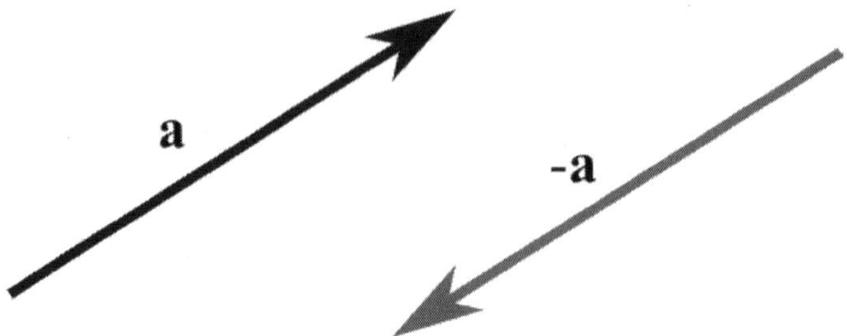

We define subtraction as addition with the opposite of a vector:

b–a=b+(–a).

This is equivalent to turning vector a around in the applying the above rules for addition. Can you see how the vector x in the below figure is equal to b”a? Notice how this is the same as stating that a+x=b, just like with subtraction of scalar numbers.

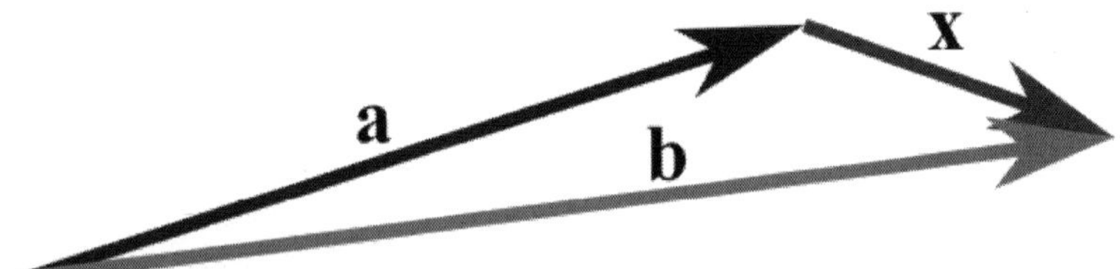

Scalar Multiplication

Given a vector a and a real number (scalar) λ, we can form the vector λa as follows. If λ is positive, then »a is the vector whose direction is the same as the direction of a and whose length is λ times the length of a. In this case, multiplication by » simply stretches (if $\lambda>1$) or compresses (if $0<\lambda<1$) the vector a.

If, on the other hand, λ is negative, then we have to take the opposite of a before stretching or compressing it. In other words, the vector λa points in the opposite direction of a, and the length of »a is $|\lambda|$ times the length of a. No matter the sign of λ, we observe that the magnitude of λa is $|\lambda|$ times the magnitude of a: $\| \lambda \mathbf{a} \| = | \lambda | \| \mathbf{a} \|$.

Scalar multiplications satisfies many of the same properties as the usual multiplication.

1. s(a+b)=sa+sb (distributive law, form 1)
2. (s+t)a=sa+ta (distributive law, form 2)

3. $1a=a$
4. $(-1)a=-a$
5. $0a=0$

In the last formula, the zero on the left is the number 0, while the zero on the right is the vector 0, which is the unique vector whose length is zero.

If $a=\lambda b$ for some scalar λ, then we say that the vectors a and b are parallel. If λ is negative, some people say that a and b are anti-parallel, but we will not use that language.

Concept of Vector Analysis

Vector analysis (or vector calculus) is a branch of mathematics concerned with differentiation and integration of vector fields, primarily in 3-dimensional Euclidean space $\mathbb{R}^3$. The term "vector calculus" is sometimes used as a synonym for the broader subject of multivariable calculus, which includes vector calculus as well as partial differentiation and multiple integration. Vector calculus plays an important role in differential geometry and in the study of partial differential equations. It is used extensively in physics and engineering, especially in the description of electromagnetic fields, gravitational fields and fluid flow.

Vector calculus was developed from quaternion analysis by J. Willard Gibbs and Oliver Heaviside near the end of the 19th century, and most of the notation and terminology was established by Gibbs and Edwin Bidwell Wilson in their 1901 book, Vector Analysis. In the conventional form using cross products, vector calculus does not generalize to higher dimensions, while the alternative approach of geometric algebra, which uses exterior products does generalize.

Basic Objects

The basic objects in vector calculus are scalar fields (scalar-valued functions) and vector fields (vector-valued functions). These are then combined or transformed under various operations, and integrated. In more advanced treatments, one further distinguishes pseudovector fields and pseudoscalar fields, which are identical to vector fields and scalar fields except that they change sign under an orientation-reversing map: for example, the curl of a vector field is a pseudovector field, and if one reflects a vector field, the curl points in the opposite direction. This distinction is clarified and elaborated in geometric algebra, as described below.

Vector Operations

Algebraic Operations: The basic algebraic (non-differential) operations in vector calculus are referred to as vector algebra, being defined for a vector space and then globally applied to a vector field, and consist of:

Scalar Multiplication: multiplication of a scalar field and a vector field, yielding a vector field: $a\mathbf{v}$;

Vector Addition: addition of two vector fields, yielding a vector field: $\mathbf{v}_1 + \mathbf{v}_2$;

Dot Product: multiplication of two vector fields, yielding a scalar field: $\mathbf{v}_1 \cdot \mathbf{v}_2$;

Cross Product: multiplication of two vector fields, yielding a vector field: $\mathbf{v}_1 \times \mathbf{v}_2$;

There are also Two Triple Products

Scalar Triple Product: the dot product of a vector and a cross product of two vectors: $\mathbf{v}_1 \cdot (\mathbf{v}_2 \times \mathbf{v}_3)$;

Vector Triple Product: the cross product of a vector and a cross product of two vectors: $\mathbf{v}_1 \times (\mathbf{v}_2 \times \mathbf{v}_3)$ or $(\mathbf{v}_3 \times \mathbf{v}_2) \times \mathbf{v}_1$; although these are less often used as basic operations, as they can be expressed in terms of the dot and cross products.

Differential Operations

Vector calculus studies various differential operators defined on scalar or vector fields, which are typically expressed in terms of the del operator (∇), also known as "nabla". The five most important differential operations in vector calculus are:

Operation	***Notation***	***Description***	***Domain/Range***
Gradient	$\text{grad}(f) = \nabla f$	Measures the rate and direction of change in a scalar field.	Maps scalar fields to vector fields.
Curl	$\text{curl}(\mathbf{F}) = \nabla \times \mathbf{F}$	Measures the tendency to rotate about a point in a vector field.	Maps vector fields to (pseudo)vector fields.
Divergence	$\text{div}(\mathbf{F}) = \nabla \cdot \mathbf{F}$	Measures the scalar of a source or sink at a given point in a vector field.	Maps vector fields to scalar fields.
Vector Laplacian	$\nabla^2 \mathbf{F} = \nabla(\nabla \cdot \mathbf{F}) - \nabla \times (\nabla \times \mathbf{F})$	Measures the difference between the value of the vector field with its average on infinitesimal balls.	Maps between vector fields.
Laplacian	$\Delta f = \nabla^2 f = \nabla \cdot \nabla f$	Measures the difference between the value of the scalar field with its average on infinitesimal balls.	Maps between scalar fields.

where the curl and divergence differ because the former uses a cross product and the latter a dot product, f denotes a scalar field and F denotes a vector field. A quantity called the Jacobian is useful for studying functions when both the domain and range of the function are multivariable, such as a change of variables during integration.

Theorems

Likewise, there are several important theorems related to these operators which generalize the fundamental theorem of calculus to higher dimensions:

Theorem	*Statement*	*Description*
Gradient theorem	$\int_{L[\mathbf{p}\to\mathbf{q}]\subset\mathbb{R}^n} \nabla\varphi\cdot d\mathbf{r} = \varphi(\mathbf{q}) - \varphi(\mathbf{p})$	The line integral through a gradient (vector) field equals the difference in its scalar field at the endpoints of the curve L.
Green's theorem	$\iint_{A\subset\mathbb{R}^2} \left(\frac{\partial M}{\partial x} - \frac{\partial L}{\partial y}\right) d\mathbf{A} = \oint_{\partial A} (L\,dx + M\,dy)$	The integral of the scalar curl of a vector field over some region in the plane equals the line integral of the vector field over the closed curve bounding the region oriented in the counterclockwise direction.
Stokes' theorem	$\iint_{\Sigma\subset\mathbb{R}^3} \nabla\times\mathbf{F}\cdot d\mathbf{\Sigma} = \oint_{\partial\Sigma} \mathbf{F}\cdot d\mathbf{r}$	The integral of the curl of a vector field over a surface in $\mathbb{R}^3$ equals the line integral of the vector field over the closed curve bounding the surface.
Divergence theorem	$\iiint_{V\subset\mathbb{R}^3} (\nabla\cdot\mathbf{F})\,d\mathbf{V} = ∎\partial V\,\mathbf{F}\cdot d\mathbf{S}$	The integral of the divergence of a vector field over some solid equals the integral of the flux through the closed surface bounding the solid.

Generalizations

Different 3-Manifolds: Vector calculus is initially defined for Euclidean 3-space, $\mathbb{R}^3$, which has additional structure beyond simply being a 3-dimensional real vector space, namely: an inner product (the dot product), which gives a notion of length (and hence angle), and an orientation, which gives a notion of left-handed and right-handed. These structures give rise to a volume form, and also the cross product, which is used pervasively in vector calculus.

The gradient and divergence require only the inner product, while the curl and the cross product also requires the handedness of the coordinate system to be taken into account.

Vector calculus can be defined on other 3-dimensional real vector spaces if they have an inner product (or more generally a symmetric nondegenerate form) and an orientation; note that this is less data than an isomorphism to Euclidean space, as it does not require a set of coordinates (a frame of reference), which reflects the fact that vector calculus is invariant under rotations (the special orthogonal group SO(3)).

More generally, vector calculus can be defined on any 3-dimensional oriented Riemannian manifold, or more generally pseudo-Riemannian manifold. This structure simply means that the tangent space at each point has an inner product (more generally, a symmetric nondegenerate form) and an orientation, or more globally that there is a symmetric nondegenerate metric tensor and an orientation, and works because vector calculus is defined in terms of tangent vectors at each point.

Other Dimensions

Most of the analytic results are easily understood, in a more general form, using the machinery of differential geometry, of which vector calculus forms a subset. Grad and div generalize immediately to other dimensions, as do the gradient theorem, divergence theorem, and Laplacian (yielding harmonic analysis), while curl and cross product do not generalize as directly.

From a general point of view, the various fields in (3-dimensional) vector calculus are uniformly seen as being k-vector fields: scalar fields are 0-vector fields, vector fields are 1-vector fields, pseudovector fields are 2-vector fields, and pseudoscalar fields are 3-vector fields. In higher dimensions there are additional types of fields (scalar/vector/pseudovector/pseudoscalar corresponding to $0/1/n{-}1/n$ dimensions, which is exhaustive in dimension 3), so one cannot only work with (pseudo)scalars and (pseudo)vectors.

In any dimension, assuming a nondegenerate form, grad of a scalar function is a vector field, and div of a vector field is a scalar function, but only in dimension 3 and 7 [1] (and, trivially, dimension 0) is the curl of a vector field a vector field, and only in 3 or 7 dimensions can a cross product be defined (generalizations in other dimensionalities either require $n-1$ vectors to yield 1 vector, or are alternative Lie algebras, which are more general antisymmetric bilinear products). The generalization of grad and div, and how curl may be generalized is elaborated at Curl: Generalizations; in brief, the curl of a vector field is a bivector field, which may be interpreted as the special orthogonal Lie algebra of infinitesimal rotations; however, this cannot be identified with a vector field because the dimensions differ - there are 3 dimensions of rotations in 3 dimensions, but 6 dimensions of rotations in 4 dimensions (and more generally $\binom{n}{2} = \frac{1}{2}n(n-1)$ dimensions of rotations in n dimensions).

There are two important alternative generalizations of vector calculus. The first, geometric algebra, uses k-vector fields instead of vector fields (in 3 or fewer dimensions, every k-vector field can be identified with a scalar function or vector field, but this is not true in higher dimensions).

This replaces the cross product, which is specific to 3 dimensions, taking in two vector fields and giving as output a vector field, with the exterior product, which exists in all dimensions and takes in two vector fields, giving as output a bivector (2-vector) field. This product yields Clifford algebras as the algebraic structure on vector spaces (with an orientation and nondegenerate form). Geometric algebra is mostly used in generalizations of physics and other applied fields to higher dimensions.

The second generalization uses differential forms (k-covector fields) instead of vector fields or k-vector fields, and is widely used in mathematics, particularly in differential geometry, geometric topology, and harmonic analysis, in particular yielding Hodge theory on oriented pseudo-Riemannian manifolds. From this point of view, grad, curl, and div correspond to the exterior derivative of 0-forms, 1-forms, and 2-forms, respectively, and the key theorems of vector calculus are all special cases of the general form of Stokes' theorem.

From the point of view of both of these generalizations, vector calculus implicitly identifies mathematically distinct objects, which makes the presentation simpler but the underlying mathematical structure and generalizations less clear. From the point of view of geometric algebra, vector calculus implicitly identifies k-vector fields with vector fields or scalar functions: 0-vectors and 3-vectors with scalars, 1-vectors and 2-vectors with vectors. From the point of view of differential forms, vector calculus implicitly identifies k-forms with scalar fields or vector fields: 0-forms and 3-forms with scalar fields, 1-forms and 2-forms with vector fields. Thus for example the curl naturally takes as input a vector field, but naturally has as output a 2-vector field or 2-form (hence pseudovector field), which is then interpreted as a vector field, rather than directly taking a vector field to a vector field; this is reflected in the curl of a vector field in higher dimensions not having as output a vector field.

Vector-Valued Function

A vector-valued function, also referred to as a vector function, is a mathematical function of one or more variables whose range is a set of multidimensional vectors or infinite-dimensional vectors. The input of a vector-valued function could be a scalar or a vector. The dimension of the domain is not defined by the dimension of the range.

Example

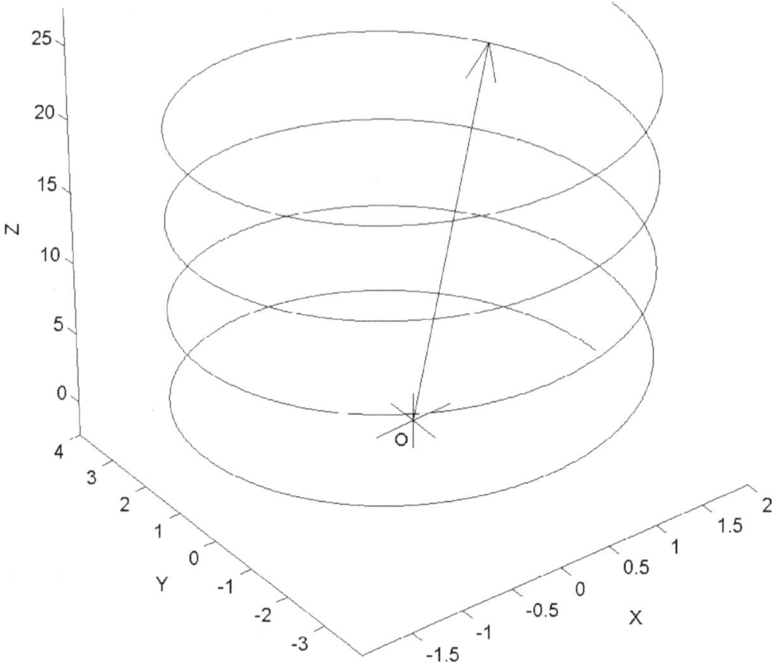

Figure: *A graph of the vector-valued function r(t) = <2 cos t, 4 sin t, t> indicating a range of solutions and the vector when evaluated near t = 19.5*

A common example of a vector valued function is one that depends on a single real number parameter *t*, often representing time, producing a vector v(*t*) as the result. In terms of the standard unit vectors i, j, k of Cartesian 3-space, these specific type of vector-valued functions are given by expressions such as

- $\mathbf{r}(t) = f(t)\mathbf{i} + g(t)\mathbf{j}$ or
- $\mathbf{r}(t) = f(t)\mathbf{i} + g(t)\mathbf{j} + h(t)\mathbf{k}$

where *f*(*t*), *g*(*t*) and *h*(*t*) are the coordinate functions of the parameter *t*. The vector r(*t*) has its tail at the origin and its head at the coordinates evaluated by the function.

The vector shown in the graph to the right is the evaluation of the function near *t*=19.5 (between 6π and 6.5π; i.e., somewhat more than 3 rotations). The spiral is the path traced by the tip of the vector as t increases from zero through 8π.

Vector functions can also be referred to in a different notation:

- $\mathbf{r}(t) = \langle f(t), g(t)\rangle$ or
- $\mathbf{r}(t) = \langle f(t), g(t), h(t)\rangle$

Properties

The domain of a vector-valued function is the intersection of the domain of the functions *f*, *g*, and *h*.

Derivative of a Three-Dimensional Vector Function

Many vector-valued functions, like scalar-valued functions, can be differentiated by simply differentiating the components in the Cartesian coordinate system. Thus, if

$$\mathbf{r}(t) = f(t)\mathbf{i} + g(t)\mathbf{j} + h(t)\mathbf{k}$$

is a vector-valued function, then

$$\frac{d\mathbf{r}(t)}{dt} = f'(t)\mathbf{i} + g'(t)\mathbf{j} + h'(t)\mathbf{k}.$$

The vector derivative admits the following physical interpretation: if r(*t*) represents the position of a particle, then the derivative is the velocity of the particle

$$\mathbf{v}(t) = \frac{d\mathbf{r}(t)}{dt}.$$

Likewise, the derivative of the velocity is the acceleration

$$\frac{d\mathbf{v}(t)}{dt} = \mathbf{a}(t).$$

Partial Derivative

The partial derivative of a vector function a with respect to a scalar variable *q* is defined as

$$\frac{\partial \mathbf{a}}{\partial q} = \sum_{i=1}^{n} \frac{\partial a_i}{\partial q}\mathbf{e}_i$$

where a_i is the *scalar component* of a in the direction of e_i. It is also called the direction cosine of a and e_i or their dot product. The vectors e_1,e_2,e_3 form an orthonormal basis fixed in the reference frame in which the derivative is being taken.

Ordinary Derivative

If a is regarded as a vector function of a single scalar variable, such as time *t*, then the equation above reduces to the first ordinary time derivative of a with respect to *t*,

$$\frac{d\mathbf{a}}{dt} = \sum_{i=1}^{3} \frac{da_i}{dt}\mathbf{e}_i.$$

Total Derivative

If the vector a is a function of a number n of scalar variables q_r ($r = 1,...,n$), and each q_r is only a function of time t, then the ordinary derivative of a with respect to t can be expressed, in a form known as the total derivative, as[1]

$$\frac{d\mathbf{a}}{dt} = \sum_{r=1}^{n} \frac{\partial \mathbf{a}}{\partial q_r} \frac{dq_r}{dt} + \frac{\partial \mathbf{a}}{\partial t}.$$

Some authors prefer to use capital D to indicate the total derivative operator, as in D/Dt. The total derivative differs from the partial time derivative in that the total derivative accounts for changes in a due to the time variance of the variables q_r.

Reference Frames

Whereas for scalar-valued functions there is only a single possible reference frame, to take the derivative of a vector-valued function requires the choice of a reference frame (at least when a fixed Cartesian coordinate system is not implied as such). Once a reference frame has been chosen, the derivative of a vector-valued function can be computed using techniques similar to those for computing derivatives of scalar-valued functions. A different choice of reference frame will, in general, produce a different derivative function. The derivative functions in different reference frames have a specific kinematical relationship.

Derivative of a Vector Function with Nonfixed Bases

The above formulas for the derivative of a vector function rely on the assumption that the basis vectors e_1,e_2,e_3 are constant, that is, fixed in the reference frame in which the derivative of a is being taken, and therefore the e_1,e_2,e_3 each has a derivative of identically zero. This often holds true for problems dealing with vector fields in a fixed coordinate system, or for simple problems in physics. However, many complex problems involve the derivative of a vector function in multiple moving reference frames, which means that the basis vectors will not necessarily be constant. In such a case where the basis vectors e_1,e_2,e_3 are fixed in reference frame E, but not in reference frame N, the more general formula for the ordinary time derivative of a vector in reference frame N is

$$\frac{{}^{\mathrm{N}}d\mathbf{a}}{dt} = \sum_{i=1}^{3} \frac{da_i}{dt}\mathbf{e}_i + \sum_{i=1}^{3} a_i \frac{{}^{\mathrm{N}}d\mathbf{e}_i}{dt}$$

where the superscript N to the left of the derivative operator indicates the reference frame in which the derivative is taken. As shown previously, the first term on the right hand side is equal to the derivative of a in the reference frame where e_1, e_2, e_3 are constant, reference frame E. It also can be shown that the second term on the right hand side is equal to the relative angular velocity of the two reference frames cross multiplied with the vector a itself. Thus, after substitution, the formula relating the derivative of a vector function in two reference frames is

$$\frac{{}^{\mathrm{N}}d\mathbf{a}}{dt} = \frac{{}^{\mathrm{E}}d\mathbf{a}}{dt} + {}^{\mathrm{N}}\boldsymbol{\omega}^{\mathrm{E}} \times \mathbf{a}$$

where ${}^{N}\omega^{E}$ is the angular velocity of the reference frame E relative to the reference frame N.

One common example where this formula is used is to find the velocity of a space-borne object, such as a rocket, in the inertial reference frame using measurements of the rocket's velocity relative to the ground. The velocity ${}^{N}v^{R}$ in inertial reference frame N of a rocket R located at position r^R can be found using the formula

$$\frac{{}^{\mathrm{N}}d}{dt}(\mathbf{r}^{\mathrm{R}}) = \frac{{}^{\mathrm{E}}d}{dt}(\mathbf{r}^{\mathrm{R}}) + {}^{\mathrm{N}}\boldsymbol{\omega}^{\mathrm{E}} \times \mathbf{r}^{\mathrm{R}}.$$

where ${}^{N}\omega^{E}$ is the angular velocity of the Earth relative to the inertial frame N. Since velocity is the derivative of position, ${}^{N}v^{R}$ and ${}^{E}v^{R}$ are the derivatives of r^R in reference frames N and E, respectively. By substitution,

$${}^{\mathrm{N}}\mathbf{v}^{\mathrm{R}} = {}^{\mathrm{E}}\mathbf{v}^{\mathrm{R}} + {}^{\mathrm{N}}\boldsymbol{\omega}^{\mathrm{E}} \times \mathbf{r}^{\mathrm{R}}$$

where ${}^{E}v^{R}$ is the velocity vector of the rocket as measured from a reference frame E that is fixed to the Earth.

Derivative and Vector Multiplication

The derivative of the products of vector functions behaves similarly to the derivative of the products of scalar functions. Specifically, in the case of scalar multiplication of a vector, if p is a scalar variable function of q,

$$\frac{\partial}{\partial q}(p\mathbf{a}) = \frac{\partial p}{\partial q}\mathbf{a} + p\frac{\partial \mathbf{a}}{\partial q}.$$

In the case of dot multiplication, for two vectors a and b that are both functions of q,

$$\frac{\partial}{\partial q}(\mathbf{a}\cdot\mathbf{b}) = \frac{\partial \mathbf{a}}{\partial q}\cdot\mathbf{b} + \mathbf{a}\cdot\frac{\partial \mathbf{b}}{\partial q}.$$

Similarly, the derivative of the cross product of two vector functions is

$$\frac{\partial}{\partial q}(\mathbf{a}\times\mathbf{b}) = \frac{\partial \mathbf{a}}{\partial q}\times\mathbf{b} + \mathbf{a}\times\frac{\partial \mathbf{b}}{\partial q}.$$

Derivative of an n-Dimensional Vector Function

A function *f* of a real number *t* with values in the space R^n can be written as $f(t) = (f_1(t), f_2(t), \ldots, f_n(t))$. Its derivative equals

$$f'(t) = (f_{1'}(t), f_{2'}(t), \ldots, f_{n'}(t)).$$

If *f* is a function of several variables, say of $t \in R^m$, then the partial derivatives of the components of *f* form a $n \times m$ matrix called the Jacobian matrix *of f*.

Infinite-Dimensional Vector Functions

If the values of a function *f* lie in an infinite-dimensional vector space *X*, such as a Hilbert space, then *f* may be called an *infinite-dimensional vector function.*

If the argument of *f* is a real number and *X* is a Hilbert space, then the derivative of *f* at a point *t* can be defined as in the finite-dimensional case:

$$f'(t) = \lim_{h\to 0}\frac{f(t+h) - f(t)}{h}.$$

Most results of the finite-dimensional case also hold in the infinite-dimensional case too, mutatis mutandis. Differentiation can also be defined to functions of several variables (e.g., $t \in R^n$ or even $t \in Y$, where *Y* is an infinite-dimensional vector space).

N.B. If *X* is a Hilbert space, then one can easily show that any derivative (and any other limit) can be computed componentwise: if

$$f = (f_1, f_2, f_3, \ldots)$$

(i.e., $f = f_1e_1 + f_2e_2 + f_3e_3 + \cdots$, where $e_1, e_2, e_3, \ldots$ is an orthonormal basis of the space *X*), and $f'(t)$ exists, then

$$f'(t) = (f_{1'}(t), f_{2'}(t), f_{3'}(t), \ldots).$$

However, the existence of a componentwise derivative does not guarantee the existence of a derivative, as componentwise convergence in a Hilbert space does not guarantee convergence with respect to the actual topology of the Hilbert space.

Other Infinite-Dimensional Vector Spaces

Most of the above hold for other topological vector spaces X too. However, not as many classical results hold in the Banach space setting, e.g., an absolutely continuous function with values in a suitable Banach space need not have a derivative anywhere. Moreover, in most Banach spaces setting there are no orthonormal bases.

Infinite-dimensional vector function refers to a function whose values lie in an infinite-dimensional vector space, such as a Hilbert space or a Banach space.

Such functions are applied in most sciences including physics.

Example

Set $f_k(t) = t / k^2$ for every positive integer k and every real number t. Then values of the function

$$f(t) = (f_1(t), f_2(t), f_3(t), \ldots)$$

lie in the infinite-dimensional vector space X (or $\mathbf{R}^{\mathrm{N}}$) of real-valued sequences. For example,

$$f(2) = \left(2, \frac{2}{4}, \frac{2}{9}, \frac{2}{16}, \frac{2}{25}, \ldots\right).$$

As a number of different topologies can be defined on the space X, we cannot talk about the derivative of f without first defining the topology of X or the concept of a limit in X.

Moreover, for any set A, there exist infinite-dimensional vector spaces having the (Hamel) dimension of the cardinality of A (e.g., the space of functions $A \to K$ with finitely-many nonzero elements, where K is the desired field of scalars). Furthermore, the argument t could lie in any set instead of the set of real numbers.

Integral and Derivative

If, e.g., $f : [0,1] \to X$, where X is a Banach space or another topological vector space, the derivative of f can be defined in the standard way: $f'(t) := \lim_{h \to 0} \dfrac{f(t+h) - f(t)}{h}$.

The measurability of f can be defined by a number of ways, most important of which are Bochner measurability and weak measurability.

The most important integrals of f are called Bochner integral (when X is a Banach space) and Pettis integral (when X is a topological vector space). Both these integrals commute with linear functionals. Also L^p spaces have been defined for such functions.

Most theorems on integration and differentiation of scalar functions can be generalized to vector-valued functions, often using essentially the same proofs. Perhaps the most important exception is that absolutely continuous functions need not equal the integrals of their (a.e.) derivatives (unless, e.g., X is a Hilbert space).

Derivative

Functions with Values in a Hilbert Space: If f is a function of real numbers with values in a Hilbert space X, then the derivative of f at a point t can be defined as in the finite-dimensional case:

$$f'(t) = \lim_{h \to 0} \frac{f(t+h) - f(t)}{h}.$$

Most results of the finite-dimensional case also hold in the infinite-dimensional case too, mutatis mutandis. Differentiation can also be defined to functions of several variables (e.g., $t \in R^n$ or even $t \in Y$, where Y is an infinite-dimensional vector space).

N.B. If X is a Hilbert space, then one can easily show that any derivative (and any other limit) can be computed componentwise: if

$$f = (f_1, f_2, f_3, \ldots)$$

(i.e., $f = f_1 e_1 + f_2 e_2 + f_3 e_3 + \cdots$, where $e_1, e_2, e_3, \ldots$ is an orthonormal basis of the space X), and $f'(t)$ exists, then

$$f'(t) = (f_{1'}(t), f_{2'}(t), f_{3'}(t), \ldots).$$

However, the existence of a componentwise derivative does not guarantee the existence of a derivative, as componentwise convergence in a Hilbert space does not guarantee convergence with respect to the actual topology of the Hilbert space.

Other Infinite-Dimensional Vector Spaces

Most of the above hold for other topological vector spaces X too. However, not as many classical results hold in the Banach space setting, e.g., an absolutely continuous function with values in a suitable Banach space need not have a derivative anywhere. Moreover, in most Banach spaces setting there are no orthonormal bases.

Differential Operators

In mathematics, a differential operator is an operator defined as a function of the differentiation operator. It is helpful, as a matter of notation first, to consider differentiation as an abstract operation that accepts a function and returns another function (in the style of a higher-order function in computer science).

Notations

The most common differential operator is the action of taking the derivative itself. Common notations for taking the first derivative with respect to a variable x include:

$$\frac{d}{dx}, D, D_x, \text{ and } \partial_x.$$

When taking higher, nth order derivatives, the operator may also be written:

$$\frac{d^n}{dx^n}, \; D^n, \text{ or } D_x^n.$$

The derivative of a function f of an argument x is sometimes given as either of the following:

$$[f(x)]'$$
$$f'(x).$$

The D notation's use and creation is credited to Oliver Heaviside, who considered differential operators of the form

$$\sum_{k=0}^{n} c_k D^k$$

in his study of differential equations.

One of the most frequently seen differential operators is the Laplacian operator, defined by

$$\Delta = \nabla^2 = \sum_{k=1}^{n} \frac{\partial^2}{\partial x_k^2}.$$

Another differential operator is the Θ operator, or theta operator, defined by

$$\Theta = z\frac{d}{dz}.$$

This is sometimes also called the homogeneity operator, because its eigenfunctions are the monomials in z:

$$\Theta(z^k) = kz^k, \quad k = 0, 1, 2, \ldots$$

In n variables the homogeneity operator is given by

$$\Theta = \sum_{k=1}^{n} x_k \frac{\partial}{\partial x_k}.$$

As in one variable, the eigenspaces of Θ are the spaces of homogeneous polynomials.

In writing, following common mathematical convention, the argument of a differential operator is usually placed on the right side of the operator itself. Sometimes an alternative notation is used: The result of applying the operator to the function on the left side of the operator and on the right side of the operator, and the difference obtained when applying the differential operator to the functions on both sides, are denoted by arrows as follows:

$$f\overleftarrow{\partial_x}g = g\cdot\partial_x f$$

$$f\overrightarrow{\partial_x}g = f\cdot\partial_x g$$

$$f\overleftrightarrow{\partial_x}g = f\cdot\partial_x g - g\cdot\partial_x f.$$

Such a bidirectional-arrow notation is frequently used for describing the probability current of quantum mechanics.

Del

The differential operator del, also called nabla operator, is an important vector differential operator. It appears frequently in physics in places like the differential form of Maxwell's Equations. In three-dimensional Cartesian coordinates, del is defined:

$$\nabla = \hat{\mathbf{x}}\frac{\partial}{\partial x} + \hat{\mathbf{y}}\frac{\partial}{\partial y} + \hat{\mathbf{z}}\frac{\partial}{\partial z}.$$

Del is used to calculate the gradient, curl, divergence, and Laplacian of various objects.

Adjoint of an Operator

Given a linear differential operator T

$$Tu = \sum_{k=0}^{n} a_k(x) D^k u$$

the adjoint of this operator is defined as the operator T^* such that

$$\langle Tu, v\rangle = \langle u, T^* v\rangle$$

where the notation $\langle\cdot,\cdot\rangle$ is used for the scalar product or inner product. This definition therefore depends on the definition of the scalar product.

Hermitian Adjoint

In mathematics, specifically in functional analysis, each bounded linear operator on a Hilbert space has a corresponding adjoint operator. Adjoints of operators generalize conjugate transposes of square matrices to (possibly) infinite-dimensional situations. If one thinks of operators on a Hilbert space as "generalized complex numbers", then the adjoint of an operator plays the role of the complex conjugate of a complex number.

The adjoint of an operator A is also sometimes called the Hermitian conjugate (after Charles Hermite) of A and is denoted by A^* or $A^\dagger$ (the latter especially when used in conjunction with the bra–ket notation).

Properties

The following properties of the Hermitian adjoint of bounded operators are immediate:

1. $A^{**} = A$ – involutiveness
2. If A is invertible, then so is A^*, with $(A^*)^{H1} = (A^{H1})^*$
3. $(A + B)^* = A^* + B^*$
4. $(\lambda A)^* = \lambda A^*$, where λ denotes the complex conjugate of the complex number λ – antilinearity (together with 3.)
5. $(AB)^* = B^*\ A^*$

If we define the operator norm of A by

$$\| A \|_{op} := \sup\{\| Ax \| : \| x \| \leq 1\}$$

then

$$\| A^* \|_{op} = \| A \|_{op} .$$

Moreover,

$$\| A^* A \|_{op} = \| A \|_{op}^2 .$$

One says that a norm that satisfies this condition behaves like a "largest value", extrapolating from the case of self-adjoint operators.

The set of bounded linear operators on a Hilbert space H together with the adjoint operation and the operator norm form the prototype of a C*-algebra.

Adjoint of Densely Defined Operators

A densely defined operator A on a Hilbert space H is a linear operator whose domain D(A) is a dense linear subspace of H and whose co-domain is H.[2] Its adjoint A^* has as domain $D(A^*)$ the set of all $y \in H$ for which there is a $\in$ H H satisfying

$$\langle Ax, y\rangle = \langle x, z\rangle \quad \text{for all } x \in D(A),$$

and $A^*(y)$ equals the z defined thus.

Properties hold with appropriate clauses about domains and codomains. For instance, the last property now states that $(AB)^*$ is an extension of B^*A^* if A, B and AB are densely defined operators.

The relationship between the image of A and the kernel of its adjoint is given by:

$$\ker A^* = (\text{im } A)^\perp$$

$$(\ker A^*)^\perp = \overline{\text{im } A}$$

Proof of the first equation:[5]

$$\begin{aligned} A^*x = 0 \quad &\Leftrightarrow \langle A^*x, y\rangle = 0 \quad \forall y \in H \\ &\Leftrightarrow \langle x, Ay\rangle = 0 \quad \forall y \in H \\ &\Leftrightarrow x \perp \text{im } A \end{aligned}$$

The second equation follows from the first by taking the orthogonal complement on both sides. Note that in general, the image need not be closed, but the kernel of a continuous operator always is.

Formal Adjoint in One Variable

In the functional space of square integrable functions, the scalar product is defined by

$$\langle f, g\rangle = \int_a^b f(x)\overline{g(x)}dx,$$

where the line over $g(x)$ denotes the complex conjugate of $g(x)$. If one moreover adds the condition that f or g vanishes for $x \to a$ and $x \to b$, one can also define the adjoint of T by

$$T^*u = \sum_{k=0}^{n} (-1)^k D^k[\overline{a_k(x)}u].$$

This formula does not explicitly depend on the definition of the scalar product. It is therefore sometimes chosen as a definition of the adjoint operator. When T^* is defined according to this formula, it is called the formal adjoint of T.

A (formally) self-adjoint operator is an operator equal to its own (formal) adjoint.

Several Variables

If Ω is a domain in R^n, and P a differential operator on Ω, then the adjoint of P is defined in L2(Ω) by duality in the analogous manner:

$$\langle f, P^* g\rangle_{L^2(\Omega)} = \langle Pf, g\rangle_{L^2(\Omega)}$$

for all smooth L^2 functions f, g. Since smooth functions are dense in L^2, this defines the adjoint on a dense subset of L^2: P^* is a densely defined operator.

Example

The Sturm–Liouville operator is a well-known example of a formal self-adjoint operator. This second-order linear differential operator L can be written in the form

$$Lu = -(pu')' + qu = -(pu'' + p'u') + qu = -pu''$$
$$-p'u' + qu = (-p)D^2u + (-p')Du + (q)u.$$

This property can be proven using the formal adjoint definition above.

$$\begin{aligned} L^*u &= (-1)^2 D^2[(-p)u] + (-1)^1 D[(-p')u] + (-1)^0 (qu) \\ &= -D^2(pu) + D(p'u) + qu \\ &= -(pu)'' + (p'u)' + qu \\ &= -p''u - 2p'u' - pu'' + p''u + p'u' + qu \\ &= -p'u' - pu'' + qu \\ &= -(pu')' + qu \\ &= Lu \end{aligned}$$

This operator is central to Sturm–Liouville theory where the eigenfunctions (analogues to eigenvectors) of this operator are considered.

Properties of Differential Operators

Differentiation is linear, i.e.,

$$D(f + g) = (Df) + (Dg)$$
$$D(af) = a(Df)$$

where f and g are functions, and a is a constant.

Any polynomial in D with function coefficients is also a differential operator. We may also compose differential operators by the rule

$$(D_1 \circ D_2)(f) = D_1(D_2(f)).$$

Some care is then required: firstly any function coefficients in the operator D_2 must be differentiable as many times as the application of D_1 requires. To get a ring of such operators we must assume derivatives of all orders of the coefficients used. Secondly, this ring will not be commutative: an operator *gD* isn't the same in general as *Dg*. In fact we have for example the relation basic in quantum mechanics:

$$Dx - xD = 1.$$

The subring of operators that are polynomials in *D* with constant coefficients is, by contrast, commutative. It can be characterised another way: it consists of the translation-invariant operators.

The differential operators also obey the shift theorem.

Several Variables

The same constructions can be carried out with partial derivatives, differentiation with respect to different variables giving rise to operators that commute.

Ring of Univariate Polynomial Differential Operators

If R is a ring, let $R < D, X >$ be the non-commutative polynomial ring over R in the variable D and X, and I the two-sided ideal generated by DX-XD-1, then the ring of univariate polynomial differential operators over R is the quotient ring $R < D, X > / I$. This is a non-commutative simple ring. Every elements can be written in a unique way as a R-linear combination of monomials of the form $X^a D^b \mod I$. It supports an analogue of the Euclidean division of polynomials.

Differential modules over $R[X]$ (for the standard derivation) can be identified with modules over $R < D, X > / I$.

Ring of Multivariate Polynomial Differential Operators

If R is a ring, let $R < D_1, \ldots, D_n, X_1, \ldots, X_n >$ be the non-commutative polynomial ring over R in the variables $D_1, \ldots, D_n, X_1, \ldots, X_n$, and I the two-sided ideal generated by the elements $D_i X_j - X_i D_j - \delta_{i,j}, D_i D_j - D_j - D_i, X_i X_j - X_j X_i$ for all $1 \le i, j \le n$ where is Kronecker delta, then the ring of univariate polynomial differential operators over R is the quotient ring $R < D_1, \ldots, D_n, X_1, \ldots, X_n > / I$.

This is a non-commutative simple ring. Every elements can be written in a unique way as a R-linear combination of monomials of the form $X_1^{a_1} \ldots X_n^{a_n} D_1^{b_1} \ldots D_n^{b_n}$.

Coordinate-Independent Description

In differential geometry and algebraic geometry it is often convenient to have a coordinate-independent description of differential operators between two vector bundles. Let E and F be two vector bundles over a differentiable manifold M. An R-linear mapping of sections $P : \Gamma(E) \to \Gamma(F)$ is said to be a kth-order linear differential operator if it factors through the jet bundle $J^k(E)$. In other words, there exists a linear mapping of vector bundles

$$i_P : J^k(E) \to F$$

such that

$$P = i_P \circ j^k$$

where j^k: $\Gamma(E) \to \Gamma(J^k(E))$ is the prolongation that associates to any section of E its k-jet.

This just means that for a given section s of E, the value of $P(s)$ at a point x H M is fully determined by the kth-order infinitesimal behaviour of s in x. In particular this implies that $P(s)(x)$ is determined by the germ of s in x, which is expressed by saying that differential operators are local. A foundational result is the Peetre theorem showing that the converse is also true: any (linear) local operator is differential.

Relation to Commutative Algebra

An equivalent, but purely algebraic description of linear differential operators is as follows: an R-linear map P is a kth-order linear differential operator, if for any $k + 1$ smooth functions $f_0, \ldots, f_k \in C^\infty(M)$ we have

$$[f_k,[f_{k-1},[\cdots[f_0,P]\cdots]]=0.$$

Here the bracket $[f,P]:\Gamma(E)\to\Gamma(F)$ is defined as the commutator

$$[f,P](s)=P(f\cdot s)-f\cdot P(s).$$

This characterization of linear differential operators shows that they are particular mappings between modules over a commutative algebra, allowing the concept to be seen as a part of commutative algebra.

Examples

- In applications to the physical sciences, operators such as the Laplace operator play a major role in setting up and solving partial differential equations.

- In differential topology the exterior derivative and Lie derivative operators have intrinsic meaning.
- In abstract algebra, the concept of a derivation allows for generalizations of differential operators which do not require the use of calculus. Frequently such generalizations are employed in algebraic geometry and commutative algebra.
- In the development of holomorphic functions of a complex variable $z = x + i\,y$, sometimes a complex function is considered to be a function of two real variables x and y. Use is made of the Wirtinger derivatives, which are partial differential operators:

$$\frac{\partial}{\partial z} = \frac{1}{2}\left(\frac{\partial}{\partial x} - i\frac{\partial}{\partial y}\right) \quad , \quad \frac{\partial}{\partial \bar{z}} = \frac{1}{2}\left(\frac{\partial}{\partial x} + i\frac{\partial}{\partial y}\right).$$

This approach is also used to study functions of several complex variables and functions of a motor variable.

Integral Transformation Theorems

In mathematics, an integral transform is any transform T of the following form:

$$(Tf)(u) = \int_{t_1}^{t_2} K(t,u)\, f(t)\, dt$$

The input of this transform is a function f, and the output is another function Tf. An integral transform is a particular kind of mathematical operator.

There are numerous useful integral transforms. Each is specified by a choice of the function K of two variables, the kernel function or nucleus of the transform.

Some kernels have an associated *inverse kernel* $K^{-1}(u, t)$ which (roughly speaking) yields an inverse transform:

$$f(t) = \int_{u_1}^{u_2} K^{-1}(u,t)(Tf(u))\, du$$

A *symmetric kernel* is one that is unchanged when the two variables are permuted.

Motivation

Mathematical notation aside, the motivation behind integral transforms is easy to understand. There are many classes of problems that are difficult to solve—or at least quite unwieldy algebraically—in their original representations. An integral transform "maps" an equation from its original "domain" into another domain. Manipulating and solving the equation in the target domain can be much easier than manipulation and solution in the original domain. The solution is then mapped back to the original domain with the inverse of the integral transform.

Also there are many applications of probability that rely on integral transforms, such as "pricing kernel" or stochastic discount factor, or the smoothing of data recovered from robust statistics.

The precursor of the transforms were the Fourier series to express functions in finite intervals. Later the Fourier transform was developed to remove the requirement of finite intervals.

Using the Fourier series, just about any practical function of time (the voltage across the terminals of an electronic device for example) can be represented as a sum of sines and cosines, each suitably scaled (multiplied by a constant factor), shifted (advanced or retarded in time) and "squeezed" or "stretched" (increasing or decreasing the frequency). The sines and cosines in the Fourier series are an example of an orthonormal basis.

Usage Example

As an example of an application of integral transforms, consider the Laplace transform. This is a technique that maps differential or integro-differential equations in the "time" domain into polynomial equations in what is termed the "complex frequency" domain. (Complex frequency is similar to actual, physical frequency but rather more general. Specifically, the imaginary component ω of the complex frequency $s = -\sigma + i\omega$ corresponds to the usual concept of frequency, *viz.*, the rate at which a sinusoid cycles, whereas the real component σ of the complex frequency corresponds to the degree of "damping".) The equation cast in terms of complex frequency is readily solved in the complex frequency domain (roots of the polynomial equations in the complex frequency domain correspond to eigenvalues in the time domain), leading to a "solution" formulated in the frequency domain. Employing the inverse transform, *i.e.*, the inverse procedure of the original Laplace transform, one obtains a time-domain solution. In this example, polynomials in the complex frequency domain (typically occurring in the denominator) correspond to power series in the time domain, while axial shifts in the complex frequency domain correspond to damping by decaying exponentials in the time domain.

The Laplace transform finds wide application in physics and particularly in electrical engineering, where the characteristic equations that describe the behaviour of an electric circuit in the complex frequency domain correspond to linear combinations of exponentially damped, scaled, and time-shifted sinusoids in the time domain. Other integral transforms find special applicability within other scientific and mathematical disciplines.

Different Domains

Here integral transforms are defined for functions on the real numbers, but they can be defined more generally for functions on a group.

- If instead one uses functions on the circle (periodic functions), integration kernels are then biperiodic functions; convolution by functions on the circle yields circular convolution.
- If one uses functions on the cyclic group of order n (C_n or $\mathbf{Z}/n\mathbf{Z}$), one obtains $n \times n$ matrices as integration kernels; convolution corresponds to circulant matrices.

General Theory

Although the properties of integral transforms vary widely, they have some properties in common. For example, every integral transform is a linear operator, since the integral is a linear operator, and in fact if the kernel is allowed to be a generalized function then all linear operators are integral transforms (a properly formulated version of this statement is the Schwartz kernel theorem).

The general theory of such integral equations is known as Fredholm theory. In this theory, the kernel is understood to be a compact operator acting on a Banach space of functions. Depending on the situation, the kernel is then variously referred to as the Fredholm operator, the nuclear operator or the Fredholm kernel.

Cartesian Coordinate System

A Cartesian coordinate system is a coordinate system that specifies each point uniquely in a plane by a pair of numerical coordinates, which are the signed distances from the point to two fixed perpendicular directed lines, measured in the same unit of length. Each reference line is called a *coordinate axis* or just *axis* of the system, and the point where they meet is its origin, usually at ordered pair (0, 0). The coordinates can also be defined as the positions of the perpendicular projections of the point onto the two axes, expressed as signed distances from the origin.

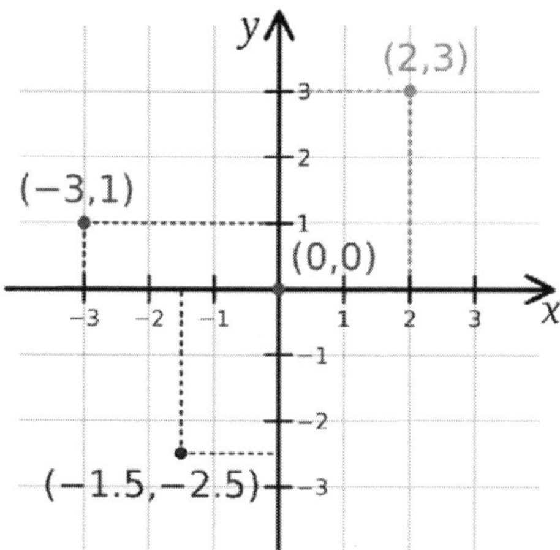

Figure: *Illustration of a Cartesian coordinate plane. Four points are marked and labelled with their coordinates: (2,/ 3) in green, (–3,/ 1) in red, (–1.5,/ 2.5) in blue, and the origin (0,/ 0) in purple.*

One can use the same principle to specify the position of any point in three-dimensional space by three Cartesian coordinates, its signed distances to three mutually perpendicular planes (or, equivalently, by its perpendicular projection onto three mutually perpendicular lines). In general, n Cartesian coordinates (an element of real n-space) specify the point in an n-dimensional Euclidean space for any dimension n. These coordinates are equal, up to sign, to distances from the point to n mutually perpendicular hyperplanes.

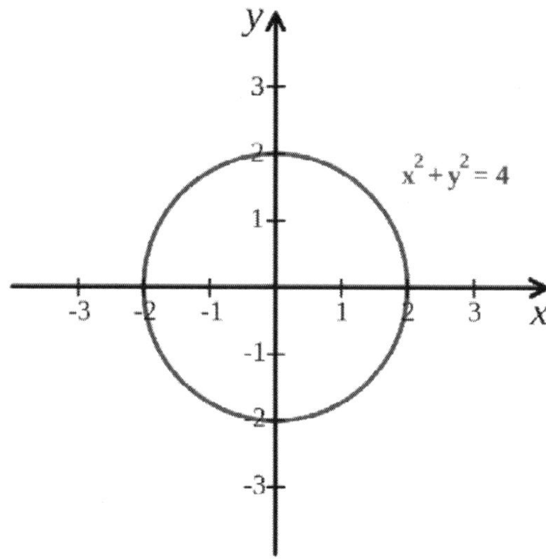

Figure: *Cartesian coordinate system with a circle of radius 2 centered at the origin. The equation of a circle is* $(x - a)^2 + (y - b)^2 = r^2$ *where a and b are the coordinates of the center (a, b) and r is the radius.*

The invention of Cartesian coordinates in the 17th century by René Descartes (Latinized name: *Cartesius*) revolutionized mathematics by providing the first systematic link between Euclidean geometry and algebra. Using the Cartesian coordinate system, geometric shapes (such as curves) can be described by Cartesian equations: algebraic equations involving the coordinates of the points lying on the shape. For example, a circle of radius 2 in a plane may be described as the set of all points whose coordinates x and y satisfy the equation $x^2 + y^2 = 4$.

Cartesian coordinates are the foundation of analytic geometry, and provide enlightening geometric interpretations for many other branches of mathematics, such as linear algebra, complex analysis, differential geometry, multivariate calculus, group theory, and more. A familiar example is the concept of the graph of a function. Cartesian coordinates are also essential tools for most applied disciplines that deal with geometry, including astronomy, physics, engineering, and many more. They are the most common coordinate system used in computer graphics, computer-aided geometric design, and other geometry-related data processing.

The adjective *Cartesian* refers to the French mathematician and philosopher René Descartes (who used the name *Cartesius* in Latin).

The idea of this system was developed in 1637 in writings by Descartes and independently by Pierre de Fermat, although Fermat also worked in three dimensions and did not publish the discovery. Both authors used a single axis in their treatments and have a variable length measured in reference to this axis. The concept of using a pair of axes was introduced later, after Descartes' La Géométrie was translated into Latin in 1649 by Frans van Schooten and his students. These commentators introduced several concepts while trying to clarify the ideas contained in Descartes' work.

The development of the Cartesian coordinate system would play an intrinsic role in the development of the calculus by Isaac Newton and Gottfried Wilhelm Leibniz.

Nicole Oresme, a French cleric and friend of the Dauphin (later to become King Charles V) of the 14th Century, used constructions similar to Cartesian coordinates well before the time of Descartes and Fermat.

Many other coordinate systems have been developed since Descartes, such as the polar coordinates for the plane, and the spherical and cylindrical coordinates for three-dimensional space.

Description

One Dimension: Choosing a Cartesian coordinate system for a one-dimensional space—that is, for a straight line—involves choosing a point O of the line (the origin), a unit of length, and an orientation for the line. An orientation chooses which of the two half-lines determined by O is the positive, and which is negative; we then say that the line "is oriented" (or "points") from the negative half towards the positive half. Then each point P of the line can be specified by its distance from O, taken with a + or – sign depending on which half-line contains P.

A line with a chosen Cartesian system is called a number line. Every real number has a unique location on the line. Conversely, every point on the line can be interpreted as a number in an ordered continuum such as the real numbers.

Two Dimensions

The modern Cartesian coordinate system in two dimensions (also called a rectangular coordinate system) is defined by an ordered pair of perpendicular lines (axes), a single unit of length for both axes, and an orientation for each axis. (Early systems allowed "oblique" axes, that is, axes that did not meet at right angles.) The lines are commonly referred to as the x- and y-axes where the x-axis is taken to be horizontal and the y-axis is taken to be vertical. The point where the axes meet is taken as the origin for both, thus turning each axis into a number line. For a given point P, a line is drawn through P perpendicular to the x-axis to meet it at X and second line is drawn through P perpendicular to the y-axis to meet it at Y. The coordinates of P are then X and Y interpreted as numbers x and y on the corresponding number lines. The coordinates are written as an ordered pair (x, y).

The point where the axes meet is the common origin of the two number lines and is simply called the *origin*. It is often labelled O and if so then the axes are called Ox and Oy. A plane with x- and y-axes defined is often referred to as the Cartesian plane or xy plane. The value of x is called the x-coordinate or abscissa and the value of y is called the y-coordinate or ordinate.

The choices of letters come from the original convention, which is to use the latter part of the alphabet to indicate unknown values. The first part of the alphabet was used to designate known values.

In the Cartesian plane, reference is sometimes made to a unit circle or a unit hyperbola.

Three Dimensions

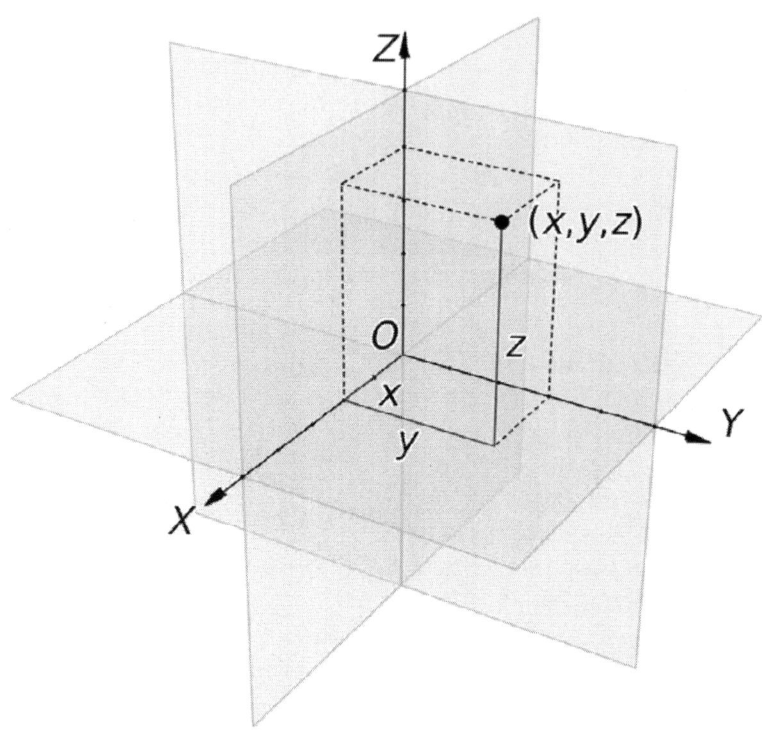

Figure: *A three dimensional Cartesian coordinate system, with origin* O *and axis lines* X, Y *and* Z, *oriented as shown by the arrows. The tick marks on the axes are one length unit apart. The black dot shows the point with coordinates* x = *2,* y = *3, and* z = *4, or (2,3,4).*

Choosing a Cartesian coordinate system for a three-dimensional space means choosing an ordered triplet of lines (axes) that are pair-wise perpendicular, have a single unit of length for all three axes and have an orientation for each axis. As in the two-dimensional case, each axis becomes a number line. The coordinates of a point *P* are obtained by drawing a line through *P* perpendicular to each coordinate axis, and reading the points where these lines meet the axes as three numbers of these number lines.

Alternatively, the coordinates of a point *P* can also be taken as the (signed) distances from *P* to the three planes defined by the three axes. If the axes are named x, y, and z, then the x-coordinate is the distance from the plane defined by the y and z axes. The distance is to be taken with the + or " sign, depending on which of the two half-spaces separated by that plane contains *P*. The y and z coordinates can be obtained in the same way from the x–z and x–y planes respectively.

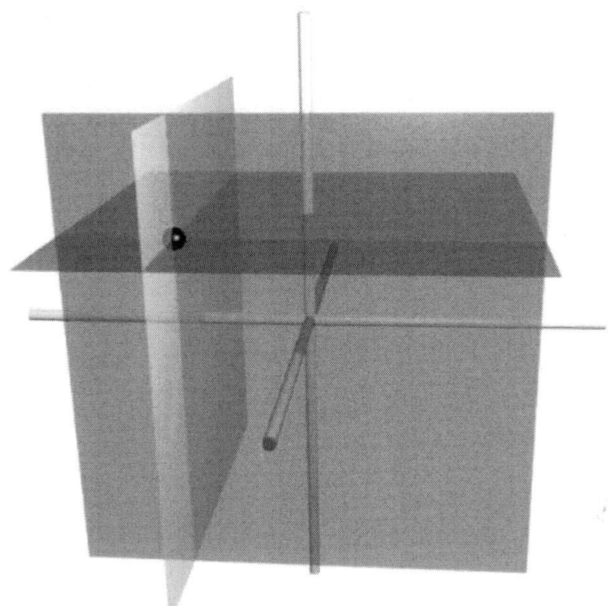

Figure: *The coordinate surfaces of the Cartesian coordinates* (x, y, z). *The* z*-axis is vertical and the* x*-axis is highlighted in green. Thus, the red plane shows the points with* x = *1, the blue plane shows the points with* z = *1, and the yellow plane shows the points with* y = *–1. The three surfaces intersect at the point* P *(shown as a black sphere) with the Cartesian coordinates (1, –1, 1).*

Higher Dimensions

A Euclidean plane with a chosen Cartesian system is called a Cartesian plane. Since Cartesian coordinates are unique and non-ambiguous, the points of a Cartesian plane can be identified with pairs of real numbers; that is with the Cartesian product $\mathbb{R}^2 = \mathbb{R} \times \mathbb{R}$, where $\mathbb{R}$ is the set of all reals. In the same way, the points any Euclidean space of dimension *n* be identified with the tuples (lists) of *n* real numbers, that is, with the Cartesian product $\mathbb{R}^n$.

Generalizations

The concept of Cartesian coordinates generalizes to allow axes that are not perpendicular to each other, and/or different units along each axis. In that case, each coordinate is obtained by projecting the point onto one axis along a direction that is parallel to the other axis (or, in general, to the hyperplane defined by all the other axes). In such an oblique coordinate system the computations of distances and angles must be modified from that in standard Cartesian systems, and many standard formulas (such as the Pythagorean formula for the distance) do not hold.

Notations and Conventions

The Cartesian coordinates of a point are usually written in parentheses and separated by commas, as in (10, 5) or (3, 5, 7). The origin is often labelled with the capital letter *O*. In analytic geometry, unknown or generic coordinates are often denoted by the letters x and y on the plane, and x, y, and z in three-dimensional space. This custom comes from a convention of algebra, which use letters near the end of the alphabet for unknown values (such as were the coordinates of points in many geometric problems), and letters near the beginning for given quantities.

These conventional names are often used in other domains, such as physics and engineering, although other letters may be used. For example, in a graph showing how a pressure varies with time, the graph coordinates may be denoted t and p. Each axis is usually named after the coordinate which is measured along it; so one says the *x-axis*, the *y-axis*, the *t-axis*, etc.

Another common convention for coordinate naming is to use subscripts, as in x_1, x_2, ... x_n for the n coordinates in an n-dimensional space; especially when n is greater than 3, or not specified. Some authors prefer the numbering x_0, x_1, ... x_{n-1}. These notations are especially advantageous in computer programming: by storing the coordinates of a point as an array, instead of a record, the subscript can serve to index the coordinates.

In mathematical illustrations of two-dimensional Cartesian systems, the first coordinate (traditionally called the abscissa) is measured along a horizontal axis, oriented from left to right. The second coordinate (the ordinate) is then measured along a vertical axis, usually oriented from bottom to top.

However, computer graphics and image processing often use a coordinate system with the y axis oriented downwards on the computer display. This convention developed in the 1960s (or earlier) from the way that images were originally stored in display buffers.

For three-dimensional systems, a convention is to portray the x–y plane horizontally, with the z axis added to represent height (positive up). Furthermore, there is a convention to orient the x-axis toward the viewer, biased either to the right or left. If a diagram (3D projection or 2D perspective drawing) shows the x and y axis horizontally and vertically, respectively, then the z axis should be shown pointing "out of the page" towards the viewer or camera. In such a 2D diagram of a 3D coordinate system, the z axis would appear as a line or ray pointing down and to the left or down and to the right, depending on the presumed viewer or camera perspective. In any diagram or display, the orientation of the three axes, as a whole, is arbitrary. However, the orientation of the axes relative to each other should always comply with the right-hand rule, unless specifically stated otherwise. All laws of physics and math assume this right-handedness, which ensures consistency.

For 3D diagrams, the names "abscissa" and "ordinate" are rarely used for x and y, respectively. When they are, the z-coordinate is sometimes called the applicate. The words *abscissa, ordinate* and *applicate* are sometimes used to refer to coordinate axes rather than the coordinate values.

Quadrants and Octants

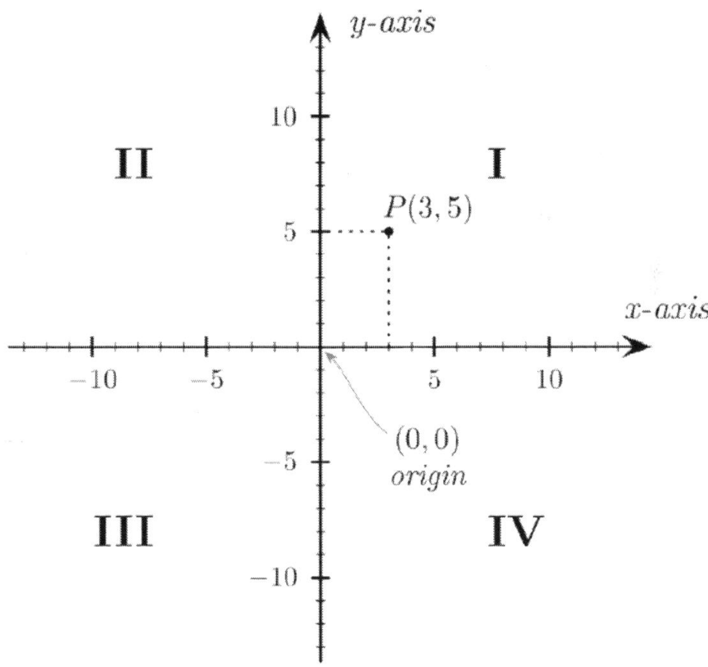

Figure: *The four quadrants of a Cartesian coordinate system.*

The axes of a two-dimensional Cartesian system divide the plane into four infinite regions, called quadrants, each bounded by two half-axes. These are often numbered from 1st to 4th and denoted by Roman numerals: I (where the signs of the two coordinates are +,+), II (–,+), III (–,–), and IV (+,–). When the axes are drawn according to the mathematical custom, the numbering goes counter-clockwise starting from the upper right ("north-east") quadrant.

Similarly, a three-dimensional Cartesian system defines a division of space into eight regions or octants, according to the signs of the coordinates of the points. The convention used for naming a specific octant is to list its signs, e.g. (+ + +) or (– + –). The generalization of the quadrant and octant to arbitrary number of dimensions is the orthant, and a similar naming system applies.

Octant (Solid Geometry)

Figure: *Three axial planes (x=0, y=0, z=0) divide space into eight equal octant domains, each with a coordinate signs from (-,-,-) to (+,+,+).*

An octant in solid geometry is one of the eight divisions of a Euclidean three-dimensional coordinate system defined by the signs of the coordinates. It is similar to the two-dimensional quadrant and the one-dimensional ray.

The generalization of an octant is called orthant.

Naming and Numbering

A convention for naming an octant is to give its list of signs, e.g. (+ - -) or (- + -). Octant (+ + +) is sometimes referred to as the *first octant*, although similar ordinal name descriptors are not defined for the other seven octants. The advantages of using the (+ - -) notation are its unambiguousness, and extensibility for higher dimensions.

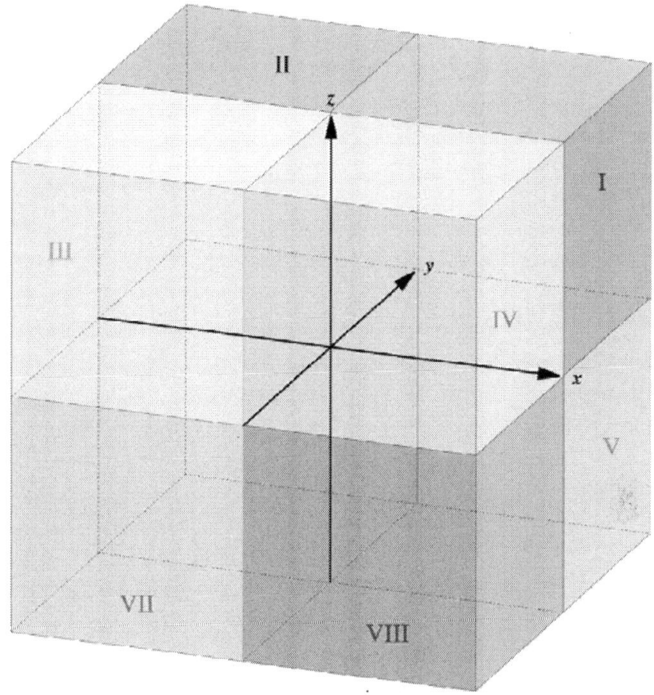

Figure: *For $z > 0$, the octants have the same numbers as the corresponding quadrants in the plane.*

Number	***Name***	x	y	z	***Octal (+=0,zyx)***	***Octal (+=1,zyx)***
I	top-front-right	+	+	+	0	7
II	top-back-right	–	+	+	1	6
III	top-back-left	–	–	+	3	4
IV	top-front-left	+	–	+	2	5
V	bottom-front-right	+	+	–	4	3
VI	bottom-back-right	–	+	–	5	2
VII	bottom-back-left	–	–	–	7	0
VIII	bottom-front-left	+	–	–	6	1

Cartesian Formulas for the Plane

Distance between Two Points

The Euclidean distance between two points of the plane with Cartesian coordinates (x_1, y_1) and (x_2, y_2) is

$$d = \sqrt{(x_2 - x_1)^2 + (y_2 - y_1)^2}.$$

This is the Cartesian version of Pythagoras' theorem. In three-dimensional space, the distance between points (x_1, y_1, z_1) and (x_2, y_2, z_2) is

$$d = \sqrt{(x_2 - x_1)^2 + (y_2 - y_1)^2 + (z_2 - z_1)^2},$$

which can be obtained by two consecutive applications of Pythagoras' theorem.

Euclidean Transformations

The Euclidean transformations or Euclidean motions are the (bijective) mappings of points of the Euclidean plane to themselves which preserve distances between points. There are four types of these mappings (also called isometries): translations, rotations, reflections and glide reflections.

Translation

Translating a set of points of the plane, preserving the distances and directions between them, is equivalent to adding a fixed pair of numbers (a, b) to the Cartesian coordinates of every point in the set. That is, if the original coordinates of a point are (x, y), after the translation they will be

$$(x', y') = (x + a, y + b).$$

Rotation

To rotate a figure counterclockwise around the origin by some angle θ is equivalent to replacing every point with coordinates (x,y) by the point with coordinates (x',y'), where

$$x' = x\cos\theta - y\sin\theta$$

$$y' = x\sin\theta + y\cos\theta.$$

Thus: $(x', y') = ((x\cos\theta - y\sin\theta), (x\sin\theta + y\cos\theta)).$

Reflection

If (x, y) are the Cartesian coordinates of a point, then $(-x, y)$ are the coordinates of its reflection across the second coordinate axis (the Y-axis), as if that line were a mirror. Likewise, $(x, -y)$ are the coordinates of its reflection across the first coordinate axis (the X-axis). In more generality, reflection across a line through the origin making an angle with the x-axis, is equivalent to replacing every point with coordinates (x, y) by the point with coordinates (x', y'), where

$$x' = x\cos 2\theta + y\sin 2\theta$$

$$y' = x\sin 2\theta - y\cos 2\theta.$$

Thus: $(x', y') = ((x\cos 2\theta + y\sin 2\theta), (x\sin 2\theta - y\cos 2\theta))$.

Glide Reflection

A glide reflection is the composition of a reflection across a line followed by a translation in the direction of that line. It can be seen that the order of these operations does not matter (the translation can come first, followed by the reflection).

General Matrix Form of the Transformations

These Euclidean transformations of the plane can all be described in a uniform way by using matrices. The result (x', y') of applying a Euclidean transformation to a point is given by (x, y) the formula

$$(x', y') = (x, y)A + b$$

where A is a 2×2 orthogonal matrix and $b = (b_1, b_2)$ is an arbitrary ordered pair of numbers;[6] that is,

$$x' = xA_{11} + yA_{21} + b_1$$

$$y' = xA_{12} + yA_{22} + b_2,$$

where

$A = \begin{pmatrix} A_{11} & A_{12} \\ A_{21} & A_{22} \end{pmatrix}$. [Note the use of row vectors for point coordinates and that the matrix is written on the right.]

To be *orthogonal*, the matrix A must have orthogonal rows with same Euclidean length of one, that is,

$$A_{11}A_{21} + A_{12}A_{22} = 0$$

and

$$A_{11}^2 + A_{12}^2 = A_{21}^2 + A_{22}^2 = 1.$$

This is equivalent to saying that A times its transpose must be the identity matrix. If these conditions do not hold, the formula describes a more general affine transformation of the plane provided that the determinant of A is not zero.

The formula defines a translation if and only if A is the identity matrix. The transformation is a rotation around some point if and only if A is a rotation matrix, meaning that

$$A_{11}A_{22} - A_{21}A_{12} = 1.$$

A reflection or glide reflection is obtained when,

$$A_{11}A_{22} - A_{21}A_{12} = -1.$$

Assuming that translation is not used transformations can be combined by simply multiplying the associated transformation matrices.

Affine Transformation

Another way to represent coordinate transformations in Cartesian coordinates is through affine transformations. In affine transformations an extra dimension is added and all points are given a value of 1 for this extra dimension. The advantage of doing this is that point translations can be specified in the final column of matrix A. In this way, all of the euclidean transformations become transactable as matrix point multiplications. The affine transformation is given by:

$$\begin{pmatrix} A_{11} & A_{21} & b_1 \\ A_{12} & A_{22} & b_2 \\ 0 & 0 & 1 \end{pmatrix} \begin{pmatrix} x \\ y \\ 1 \end{pmatrix} = \begin{pmatrix} x' \\ y' \\ 1 \end{pmatrix}.$$ [Note the matrix A from above was transposed. The matrix is on the left and column vectors for point coordinates are used.]

Using affine transformations multiple different euclidean transformations including translation can be combined by simply multiplying the corresponding matrices.

Scaling

An example of an affine transformation which is not a Euclidean motion is given by scaling. To make a figure larger or smaller is equivalent to multiplying the Cartesian coordinates of every point by the same positive number m. If (x, y) are the coordinates of a point on the original figure, the corresponding point on the scaled figure has coordinates

$$(x', y') = (mx, my).$$

If m is greater than 1, the figure becomes larger; if m is between 0 and 1, it becomes smaller.

Shearing

A shearing transformation will push the top of a square sideways to form a parallelogram. Horizontal shearing is defined by:

$$(x', y') = (x + ys, y)$$

Shearing can also be applied vertically:

Orientation and Handedness

In mathematics, orientation is a geometric notion that in two dimensions allows one to say when a cycle goes around clockwise or counterclockwise, and in three dimensions when a figure is left-handed or right-handed. In linear algebra, the notion of orientation makes sense in arbitrary dimensions. In this setting, the orientation of an ordered basis is a kind of asymmetry that makes a reflection impossible to replicate by means of a simple rotation. Thus, in three dimensions, it is impossible to make the left hand of a human figure into the right hand of the figure by applying a rotation alone, but it is possible to do so by reflecting the figure in a mirror. As a result, in the three-dimensional Euclidean space, the two possible basis orientations are called right-handed and left-handed (or right-chiral and left-chiral).

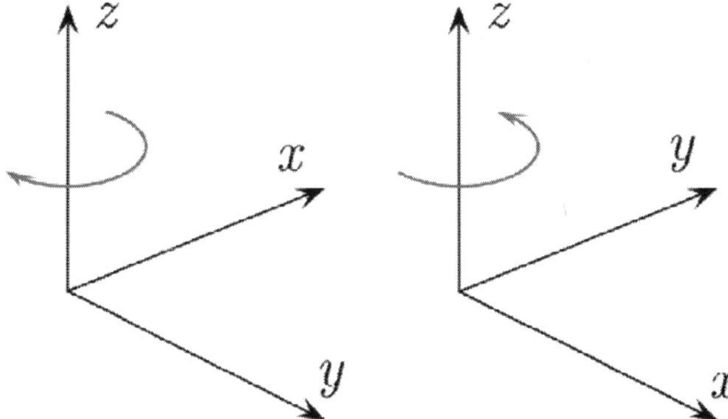

Figure: *The left-handed orientation is shown on the left, and the right-handed on the right.*

The orientation on a real vector space is the arbitrary choice of which ordered bases are "positively" oriented and which are "negatively" oriented. In the three-dimensional Euclidean space, right-handed bases are typically declared to be positively oriented, but the choice is arbitrary, as they may also be assigned a negative orientation. A vector space with an orientation selected is called an oriented vector space, while one not having an orientation selected, is called unoriented.

Definition

Let V be a finite-dimensional real vector space and let b_1 and b_2 be two ordered bases for V. It is a standard result in linear algebra that there exists a unique linear transformation $A : V \rightarrow V$ that takes b_1 to b_2. The bases b_1 and b_2 are said to have the *same orientation* (or be consistently oriented) if A has positive determinant; otherwise they have *opposite orientations*. The property of having the same orientation defines an equivalence relation on the set of all ordered bases for V. If V is non-zero, there are precisely two equivalence classes determined by this relation. An orientation on V is an assignment of +1 to one equivalence class and −1 to the other.

Every ordered basis lives in one equivalence class or another. Thus any choice of a privileged ordered basis for V determines an orientation: the orientation class of the privileged basis is declared to be positive. For example, the standard basis on R^n provides a standard orientation on R^n (in turn, the orientation of the standard basis depends on the orientation of the Cartesian coordinate system on which it is built). Any choice of a linear isomorphism between V and R^n will then provide an orientation on V.

The ordering of elements in a basis is crucial. Two bases with a different ordering will differ by some permutation. They will have the same/opposite orientations according to whether the signature of this permutation is ±1. This is because the determinant of a permutation matrix is equal to the signature of the associated permutation.

Similarly, let A be a nonsingular linear mapping of vector space R^n to R^n. This mapping is orientation-preserving if its determinant is positive. For instance, in R^3 a rotation around the Z Cartesian axis by an angle α is orientation-preserving:

$$\mathbf{A}_1 = \begin{pmatrix} \cos\alpha & -\sin\alpha & 0 \\ \sin\alpha & \cos\alpha & 0 \\ 0 & 0 & 1 \end{pmatrix}$$

while a reflection by the XY Cartesian plane is not orientation-preserving:

$$\mathbf{A}_2 = \begin{pmatrix} 1 & 0 & 0 \\ 0 & 1 & 0 \\ 0 & 0 & -1 \end{pmatrix}$$

Zero-Dimensional Case

The concept of orientation defined above did not quite apply to zero-dimensional vector spaces (as the only empty matrix is the identity (with determinant 1), so there will be only one equivalence class). However, it is useful to be able to assign different orientations to a point (e.g. orienting the boundary of a 1-dimensional manifold). A more general definition of orientation that works regardless of dimension is the following: An orientation on V is a map from the set of ordered bases of V to the set $\{\pm 1\}$ that is invariant under base changes with positive determinant and changes sign under base changes with negative determinant (it is equivarient with respect to the homomorphism $GL_n \to \pm 1$).

The set of ordered bases of the zero-dimensional vector space has one element (the empty set), and so there are two maps from this set to $\{\pm 1\}$.

A subtle point is that a zero-dimensional vector space is naturally (canonically) oriented, so we can talk about an orientation being positive (agreeing with the canonical orientation) or negative (disagreeing). An application is interpreting the Fundamental theorem of calculus as a special case of Stokes' theorem.

Two ways of seeing this are:

- A zero-dimensional vector space is a point, and there is a unique map from a point to a point, so every zero-dimensional vector space is naturally identified with R^0, and thus is oriented.
- The 0th exterior power of a vector space is the ground field K, which here is R^1, which has an orientation (given by the standard basis).

Alternate Viewpoints

Multilinear Algebra: For any n-dimensional real vector space V we can form the kth-exterior power of V, denoted $\Lambda^k V$. This is a real vector space of dimension $\binom{n}{k}$. The vector space $\Lambda^n V$ (called the *top exterior power*) therefore has dimension 1. That is, $\Lambda^n V$ is just a real line. There is no *a priori* choice of which direction on this line is positive. An orientation is just such a choice. Any nonzero linear form ω on $\Lambda^n V$ determines an orientation of V by declaring that x is in the positive direction when $\omega(x) > 0$. To connect with the basis point of view we say that the positively oriented bases are those on which ω evaluates to a positive number (since ω is an n-form we can evaluate it on an ordered set of n vectors, giving an element of R). The form ω is called an orientation form. If $\{e_i\}$ is a privileged basis for V and $\{e_i^-\}$ is the dual basis, then the orientation form giving the standard orientation is $e_1^* \wedge e_2^* \wedge \cdots \wedge e_n^*$.

The connection of this with the determinant point of view is: the determinant of an endomorphism $T: V \to V$ can be interpreted as the induced action on the top exterior power.

Lie Group Theory

Let B be the set of all ordered bases for V. Then the general linear group GL(V) acts freely and transitively on B. (In fancy language, B is a GL(V)-torsor). This means that as a manifold, B is (noncanonically) homeomorphic to GL(V). Note that the group GL(V) is not connected, but rather has two connected components according to whether the determinant of the transformation is positive or negative (except for GL_0, which is the trivial group and thus has a single connected component; this corresponds to the canonical orientation on a zero-dimensional vector space). The identity component of GL(V) is denoted $GL^+(V)$ and consists of those transformations with positive determinant. The action of $GL^+(V)$ on B is *not* transitive: there are two orbits which correspond to the connected components of B. These orbits are precisely the equivalence classes referred to above. Since B does not have a distinguished element (i.e. a privileged basis) there is no natural choice of which component is positive. Contrast this with GL(V) which does have a privileged component: the component of the identity. A specific choice of homeomorphism between B and GL(V) is equivalent to a choice of a privileged basis and therefore determines an orientation.

More formally: $\pi_0(\mathrm{GL}(V)) = (\mathrm{GL}(V) / \mathrm{GL}^+(V) = \{\pm 1\}$, and the Stiefel manifold of n-frames in is a $\mathrm{GL}(V)$ -torsor, so $V_n(V) / \mathrm{GL}^+(V)$ is a torsor over $\{\pm 1\}$, i.e., its 2 points, and a choice of one of them is an orientation.

Geometric Algebra

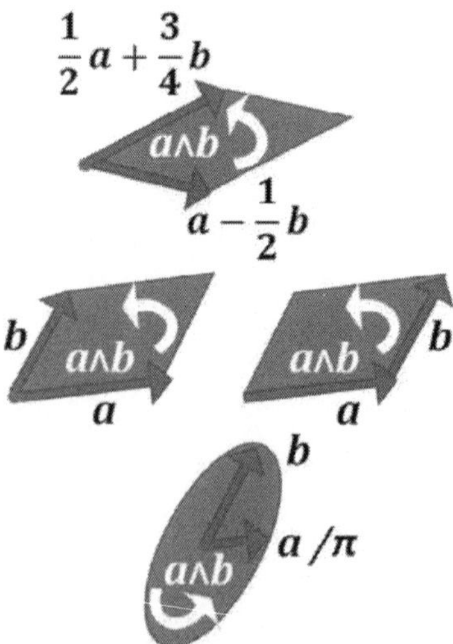

Figure: *Parallel plane segments with the same attitude, magnitude and orientation, all corresponding to the same bivector a ∧ b.[3]*

The various objects of geometric algebra are charged with three attributes or *features*: attitude, orientation, and magnitude. For example, a vector has an attitude given by a straight line parallel to it, an orientation given by its sense (often indicated by an arrowhead) and a magnitude given by its length. Similarly, a bivector in three dimensions has an attitude given by the family of planes associated with it (possibly specified by the normal line common to these planes), an orientation (sometimes denoted by a curved arrow in the plane) indicating a choice of sense of traversal of its boundary (its *circulation*), and a magnitude given by the area of the parallelogram defined by its two vectors.

Orientation on Manifolds

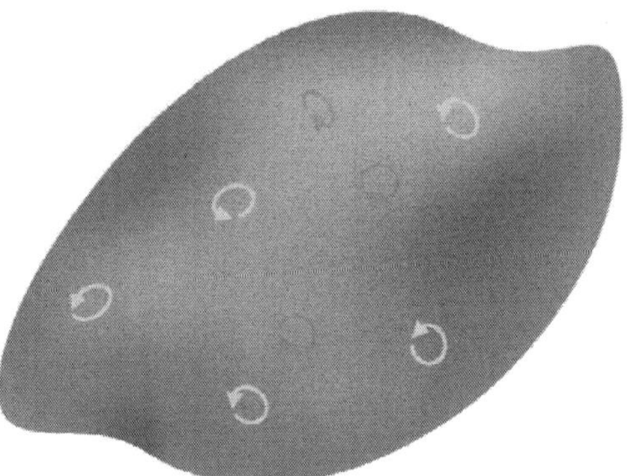

Figure: *The orientation of a volume may be determined by the orientation on its boundary, indicated by the circulating arrows.*

One can also discuss orientation on manifolds. Each point p on an n-dimensional differentiable manifold has a tangent space T_pM which is an n-dimensional real vector space. One can assign to each of these vector spaces an orientation. However, one would like to know whether it is possible to choose the orientations so that they "vary smoothly" from point to point. Due to certain topological restrictions, there are situations when this is impossible. A manifold which admits a smooth choice of orientations for its tangents spaces is said to be *orientable.*

In Two Dimensions

Fixing or choosing the x-axis determines the y-axis up to direction. Namely, the y-axis is necessarily the perpendicular to the x-axis through the point marked 0 on the x-axis. But there is a choice of which of the two half lines on the perpendicular to designate as positive and which as negative. Each of these two choices determines a different orientation (also called *handedness*) of the Cartesian plane.

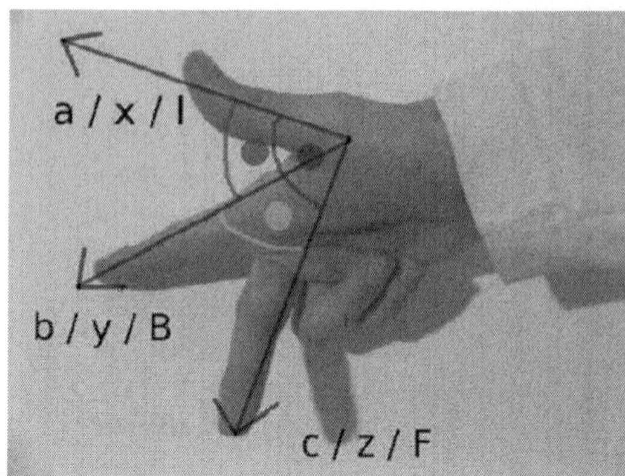

Figure: *The right hand rule.*

The usual way of orienting the axes, with the positive x-axis pointing right and the positive y-axis pointing up (and the x-axis being the "first" and the y-axis the "second" axis) is considered the *positive* or *standard* orientation, also called the *right-handed* orientation.

A commonly used mnemonic for defining the positive orientation is the right hand rule. Placing a somewhat closed right hand on the plane with the thumb pointing up, the fingers point from the x-axis to the y-axis, in a positively oriented coordinate system.

The other way of orienting the axes is following the *left hand rule*, placing the left hand on the plane with the thumb pointing up.

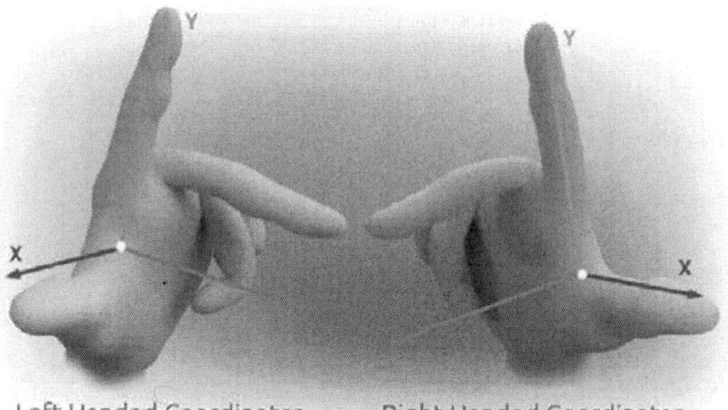

Figure: *3D Cartesian Coordinate Handedness*

When pointing the thumb away from the origin along an axis towards positive, the curvature of the fingers indicates a positive rotation along that axis.

Regardless of the rule used to orient the axes, rotating the coordinate system will preserve the orientation. Switching any two axes will reverse the orientation, but switching both will leave the orientation unchanged.

In Three Dimensions

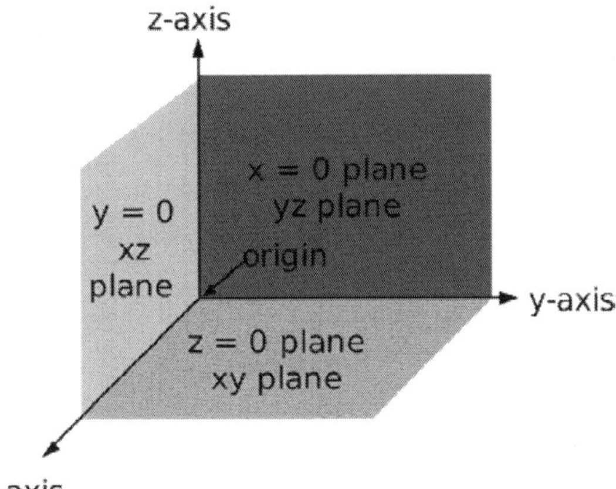

Figure: *The right-handed Cartesian coordinate system indicating the coordinate planes.*

Once the x- and y-axes are specified, they determine the line along which the z-axis should lie, but there are two possible directions on this line. The two possible coordinate systems which result are called 'right-handed' and 'left-handed'. The standard orientation, where the xy-plane is horizontal and the z-axis points up (and the x- and the y-axis form a positively oriented two-dimensional coordinate system in the xy-plane if observed from *above* the xy-plane) is called right-handed or positive.

The name derives from the right-hand rule. If the index finger of the right hand is pointed forward, the middle finger bent inward at a right angle to it, and the thumb placed at a right angle to both, the three fingers indicate the relative directions of the x-, y-, and z-axes in a *right-handed* system. The thumb indicates the x-axis, the index finger the y-axis and the middle finger the z-axis. Conversely, if the same is done with the left hand, a left-handed system results.

Figure 7 depicts a left and a right-handed coordinate system. Because a three-dimensional object is represented on the two-dimensional screen, distortion and ambiguity result. The axis pointing downward (and to the right) is also meant to point *towards* the observer, whereas the "middle" axis is meant to point *away* from the observer. The red circle is *parallel* to the horizontal xy-plane and indicates rotation from the x-axis to the y-axis (in both cases). Hence the red arrow passes *in front of* the z-axis.

Figure depicted above, is another attempt at depicting a right-handed coordinate system. Again, there is an ambiguity caused by projecting the three-dimensional coordinate system into the plane. Many observers see figure given earlier, as “flipping in and out” between a convex cube and a concave “corner”. This corresponds to the two possible orientations of the coordinate system. Seeing the figure as convex gives a left-handed coordinate system. Thus the “correct” way to view Figure above, is to imagine the x-axis as pointing *towards* the observer and thus seeing a concave corner.

Representing a Vector in the Standard Basis

A point in space in a Cartesian coordinate system may also be represented by a position vector, which can be thought of as an arrow pointing from the origin of the coordinate system to the point. If the coordinates represent spatial positions (displacements), it is common to represent the vector from the origin to the point of interest as r. In two dimensions, the vector from the origin to the point with Cartesian coordinates (x, y) can be written as:

$$\mathbf{r} = x\mathbf{i} + y\mathbf{j}$$

where $\mathbf{i} = \begin{pmatrix} 1 \\ 0 \end{pmatrix}$, and $\mathbf{j} = \begin{pmatrix} 0 \\ 1 \end{pmatrix}$ are unit vectors in the direction of the x-axis and y-axis respectively, generally referred to as the *standard basis* (in some application areas these may also be referred to as versors). Similarly, in three dimensions, the vector from the origin to the point with Cartesian coordinates (x, y, z) can be written as:

$$\mathbf{r} = x\mathbf{i} + y\mathbf{j} + z\mathbf{k}$$

where $\mathbf{k} = \begin{pmatrix} 0 \\ 0 \\ 1 \end{pmatrix}$ is the unit vector in the direction of the z-axis.

There is no *natural* interpretation of multiplying vectors to obtain another vector that works in all dimensions, however there is a way to use complex numbers to provide such a multiplication. In a two dimensional cartesian plane, identify the point with coordinates (x, y) with the complex number $z = x + iy$. Here, i is the imaginary unit and is identified with the point with coordinates (0, 1), so it is not the unit vector in the direction of the x-axis. Since the complex numbers can be multiplied giving another complex number, this identification provides a means to “multiply” vectors. In a three dimensional cartesian space a similar identification can be made with a subset of the quaternions.

Applications

Cartesian coordinates are an abstraction that have a multitude of possible applications in the real world. However, three constructive steps are involved in superimposing coordinates on a problem application. 1) Units of distance must be decided defining the spatial size represented by the numbers used as coordinates. 2) An origin must be assigned to a specific spatial location or landmark, and 3) the orientation of the axes must be defined using available directional cues for (n-1) of the n axes.

Consider as an example superimposing 3D Cartesian coordinates over all points on the Earth (i.e. geospatial 3D). What units make sense? Kilometers are a good choice, since the original definition of the kilometer was geospatial...10,000 km equalling the surface distance from Equator to North Pole. Where to place the origin? Based on symmetry, the gravitational center of the Earth suggests a natural landmark (which can be sensed via satellite orbits). Finally, how to orient X, Y and Z axis directions? The axis of Earth's spin provides a natural direction strongly associated with "up vs. down", so positive Z can adopt the direction from geocenter to North Pole. A location on the Equator is needed to define the X-axis, and the Prime Meridian stands out as a reference direction, so the X-axis takes the direction from geocenter out to [0 degrees longitude, 0 degrees latitude]. Note that with 3 dimensions, and two perpendicular axes directions pinned down for X and Z, the Y-axis is determined by the first two choices. In order to obey the right hand rule, the Y-axis must point out from the geocenter to [90 degrees longitude, 0 degrees latitude]. So what are the geocentric coordinates of the Empire State Building in New York City? Using [longitude = –73.985656, latitude = 40.748433], Earth radius = 40,000/ 2π, and transforming from spherical —> Cartesian coordinates, you can estimate the geocentric coordinates of the Empire State Building, [x, y, z] = [1330.53 km, –4635.75 km, 4155.46 km]. GPS navigation relies on such geocentric coordinates.

In engineering projects, agreement on the definition of coordinates is a crucial foundation. One cannot assume that coordinates come predefined for a novel application, so knowledge of how to erect a coordinate system where there is none is essential to applying René Descartes' ingenious thinking.

While spatial apps employ identical units along all axes, in business and scientific apps, each axis may have different units of measurement associated with it (such as kilograms, seconds, pounds, etc.). Although four- and higher-dimensional spaces are difficult to visualize, the algebra of Cartesian coordinates can be extended relatively easily to four or more variables, so that certain calculations involving many variables can be done. (This sort of algebraic extension is what is used to define the geometry of higher-dimensional spaces.) Conversely, it is often helpful to use the geometry of Cartesian coordinates in two or three dimensions to visualize algebraic relationships between two or three of many non-spatial variables.

The graph of a function or relation is the set of all points satisfying that function or relation. For a function of one variable, f, the set of all points (x, y), where $y = f(x)$ is the graph of the function f. For a function g of two variables, the set of all points (x, y, z), where $z = g(x, y)$ is the graph of the function g. A sketch of the graph of such a function or relation would consist of all the salient parts of the function or relation which would include its relative extrema, its concavity and points of inflection, any points of discontinuity and its end behaviour. All of these terms are more fully defined in calculus. Such graphs are useful in calculus to understand the nature and behaviour of a function or relation.

Chapter 2

The Conic Section

In mathematics, a conic section (or just conic) is a curve obtained as the intersection of a cone (more precisely, a right circular conical surface) with a plane. In analytic geometry, a conic may be defined as a plane algebraic curve of degree 2. There are a number of other geometric definitions possible. One of the most useful, in that it involves only the plane, is that a conic consists of those points whose distances to some point, called a focus, and some line, called a *directrix*, are in a fixed ratio, called the eccentricity.

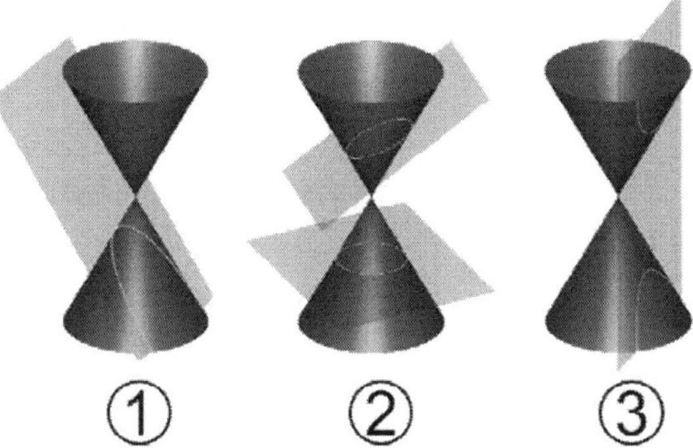

Figure: *Types of conic sections: 1. Parabola; 2. Circle and ellipse; 3. Hyperbola*

Traditionally, the three types of conic section are the hyperbola, the parabola, and the ellipse. The circle is a special case of the ellipse, and is of sufficient interest in its own right that it is sometimes called the fourth type of conic section. The type of a conic corresponds to its eccentricity, those with eccentricity less than 1 being ellipses, those with eccentricity equal to 1 being parabolas, and those with eccentricity greater than 1 being hyperbolas. In the focus-directrix definition of a conic the circle is a limiting case with eccentricity 0. In modern geometry certain degenerate cases, such as the union of two lines, are included as conics as well.

The conic sections were named and studied at least since 200 BC, when Apollonius of Perga undertook a systematic study of their properties.

Menaechmus and Early Works

It is believed that the first definition of a conic section is due to Menaechmus (died 320 BC). His work did not survive and is only known through secondary accounts. The definition used at that time differs from the one commonly used today in that it requires the plane cutting the cone to be perpendicular to one of the lines, (a generatrix), that generates the cone as a surface of revolution. Thus the shape of the conic is determined by the angle formed at the vertex of the cone (between two opposite generatrices): If the angle is acute then the conic is an ellipse; if the angle is right then the conic is a parabola; and if the angle is obtuse then the conic is a hyperbola. Note that the circle cannot be defined this way and was not considered a conic at this time.

Euclid (fl. 300 BC) is said to have written four books on conics but these were lost as well. Archimedes (died c. 212 BC) is known to have studied conics, having determined the area bounded by a parabola and an ellipse. The only part of this work to survive is a book on the solids of revolution of conics.

Apollonius of Perga

The greatest progress in the study of conics by the ancient Greeks is due to Apollonius of Perga (died c. 190 BC), whose eight-volume *Conic Sections* summarized the existing knowledge at the time and greatly extended it. Apollonius's major innovation was to characterize a conic using properties within the plane and intrinsic to the curve; this greatly simplified analysis. With this tool, it was now possible to show that any plane cutting the cone, regardless of its angle, will produce a conic according to the earlier definition, leading to the definition commonly used today.

Pappus of Alexandria (died c. 350 CE) is credited with discovering importance of the concept of a focus of a conic, and the discovery of the related concept of a directrix.

Al-Kuhi

An instrument for drawing conic sections was first described in 1000 CE by the Islamic mathematician Al-Kuhi.

Omar Khayyám

Apollonius's work was translated into Arabic (the technical language of the time) and much of his work only survives through the Arabic version. Persians found applications to the theory; the most notable of these was the Persian mathematician and poet Omar Khayyám who used conic sections to solve algebraic equations.

Europe

Johannes Kepler extended the theory of conics through the "principle of continuity", a precursor to the concept of limits.

Girard Desargues and Blaise Pascal developed a theory of conics using an early form of projective geometry and this helped to provide impetus for the study of this new field. In particular, Pascal discovered a theorem known as the hexagrammum mysticum from which many other properties of conics can be deduced.

Meanwhile, René Descartes applied his newly discovered Analytic geometry to the study of conics. This had the effect of reducing the geometrical problems of conics to problems in algebra.

Features

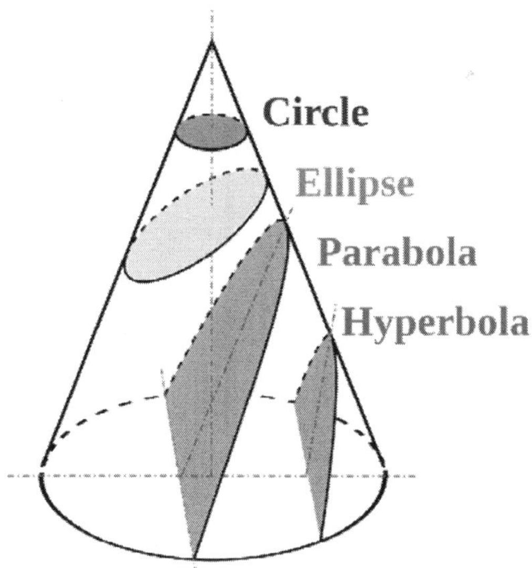

The three types of conics are the ellipse, parabola, and hyperbola. The circle can be considered as a fourth type (as it was by Apollonius) or as a kind of ellipse. The circle and the ellipse arise when the intersection of cone and plane is a closed curve. The circle is obtained when the cutting plane is parallel to the plane of the generating circle of the cone – for a right cone as in the picture at the top of the page this means that the cutting plane is perpendicular to the symmetry axis of the cone. If the cutting plane is parallel to exactly one generating line of the cone, then the conic is unbounded and is called a parabola. In the remaining case, the figure is a hyperbola. In this case, the plane will intersect *both* halves (*nappes*) of the cone, producing two separate unbounded curves.

Various parameters are associated with a conic section, as shown in the following table. (For the ellipse, the table gives the case of $a>b$, for which the major axis is horizontal; for the reverse case, interchange the symbols a and b. For the hyperbola the east-west opening case is given. In all cases, a and b are positive.)

Conic section	*Equation*	*Eccentricity (e)*	*Linear eccentricity (c)*	*Semi-latus rectum (?)*	*Focal parameter (p)*
circle	$x^2 + y^2 = a^2$	0	0	a	∞
ellipse	$\frac{x^2}{a^2} + \frac{y^2}{b^2} = 1$	$\sqrt{1 - \frac{b^2}{a^2}}$	$\sqrt{a^2 - b^2}$	$\frac{b^2}{a}$	$\frac{b^2}{\sqrt{a^2 - b^2}}$
parabola	$y^2 = 4ax$	1	a	$2a$	$2a$
hyperbola	$\frac{x^2}{a^2} - \frac{y^2}{b^2} = 1$	$\sqrt{1 + \frac{b^2}{a^2}}$	$\sqrt{a^2 + b^2}$	$\frac{b^2}{a}$	$\frac{b^2}{\sqrt{a^2 + b^2}}$

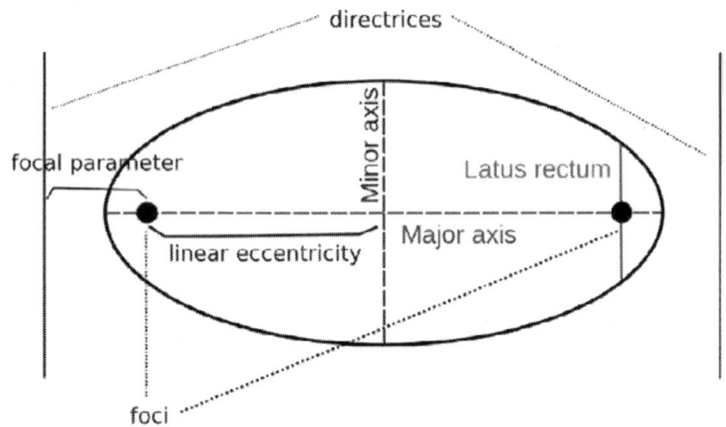

Figure: *Conic parameters in the case of an ellipse*

Conic sections are exactly those curves that, for a point F, a line L not containing F and a non-negative number e, are the locus of points whose distance to F equals e times their distance to L. F is called the focus, L the directrix, and e the eccentricity.

The linear eccentricity (c) is the distance between the center and the focus (or one of the two foci).

The latus rectum (2ℓ) is the chord parallel to the directrix and passing through the focus (or one of the two foci).

The semi-latus rectum (ℓ) is half the latus rectum.

The focal parameter (p) is the distance from the focus (or one of the two foci) to the directrix.

The following relations hold:

- $pe = \ell$
- $ae = c$.

Construction

There are many methods to construct a conic. One of them, that is useful in engineering applications, being parallelogram method, where a conic is constructed point by point by means of connecting certain equally spaced points on horizontal line and vertical line.

Properties

Just as two (distinct) points determine a line, five points determine a conic. Formally, given any five points in the plane in general linear position, meaning no three collinear, there is a unique conic passing through them, which will be non-degenerate; this is true over both the affine plane and projective plane. Indeed, given any five points there is a conic passing through them, but if three of the points are collinear the conic will be degenerate (reducible, because it contains a line), and may not be unique.

Irreducible conic sections are always “smooth”. This is important for many applications, such as aerodynamics, where a smooth surface is required to ensure laminar flow and to prevent turbulence.

Intersection at Infinity

An algebro-geometrically intrinsic form of this classification is by the intersection of the conic with the line at infinity, which gives further insight into their geometry:

- ellipses intersect the line at infinity in 0 points—rather, in 0 real points, but in 2 complex points, which are conjugate;
- parabolas intersect the line at infinity in 1 double point, corresponding to the axis—they are tangent to the line at infinity, and close at infinity, as distended ellipses;
- hyperbolas intersect the line at infinity in 2 points, corresponding to the asymptotes—hyperbolas pass through infinity, with a twist. Going to infinity along one branch passes through the point at infinity corresponding to the asymptote, then re-emerges on the other branch at the other side but with

the inside of the hyperbola (the direction of curvature) on the other side – left vs. right (corresponding to the non-orientability of the real projective plane)—and then passing through the other point at infinity returns to the first branch. Hyperbolas can thus be seen as ellipses that have been pulled through infinity and re-emerged on the other side, flipped.

Degenerate Conic

In mathematics, a degenerate conic is a conic (a second-degree plane curve, the points of which satisfy an equation that is quadratic in one or the other or both variables) that fails to be an irreducible curve. This can happen in two ways: either it is a reducible variety, meaning that its defining quadratic form is factorable as the product of two linear polynomials, or the polynomial is irreducible but defines not a curve but instead a lower-dimension variety (a point or the empty set); the latter can only occur over a field that is not algebraically closed, such as the real numbers.

In the real plane, a degenerate conic can be two lines that may or may not be parallel, a single line (actually two coinciding lines), a single point, or the null set (no points).

Examples

The conic section with equation $x^2 - y^2 = 0$ is an example of the first failure, reducibility. This conic section is degenerate because it is reducible. The equation can be written as $(x - y)(x + y) = 0$, and corresponds to two intersecting lines or an "X".

The conic section with equation $x^2 + y^2 = 0$ is an example of the second failure, not enough points (over the field of definition), over the real numbers. This conic section is degenerate because it defines only one point, $(0,0)$, not a curve. The conic section with equation $x^2 + y^2 = -1$ is likewise degenerate because it defines the empty set.

Over the field of complex numbers, the conic section with equation $x^2 + y^2 = 0$ factors as $(x + iy)(x - iy) = 0$ and is degenerate because it is reducible.

Classification

Over the complex projective plane there are only two types of degenerate conics – two different lines, which necessarily intersect in one point, or one double line.

Over the real affine plane the situation is more complicated.

Reducible

Reducible conics – those whose equation factors – consist of two lines in the plane. There are three possible configurations of these, according to how they intersect. These form a 4-dimensional space (each line has two parameters, namely a slope and a position, as is slope-intercept form), with special intersections as lower-dimensional sub-varieties.

- Two intersecting lines, a 4-dimensional space, such as $x^2 - y^2 = 0 \Leftrightarrow (x+y)(x-y) = 0$
- Two parallel lines,a 3-dimensional space, such as $x^2 - 1 = 0 \Leftrightarrow (x+1)(x-1) = 0$
- A single doubled line (multiplicity 2), a 2-dimensional space, such as $x^2 = 0$

In terms of the points at infinity, two intersecting lines have 2 distinct points at infinity, while two parallel lines intersect at 1 point at infinity (hence intersect the line at infinity in a double point), and a single double line also intersects the line at infinity in a double point.

Not Enough Points

Over a non-algebraically closed field such as the real numbers, a conic may also be degenerate because it does not have enough real points (if it has any at all). This can occur in two ways:

- A single double point, such as $x^2 + y^2 = 0$.
- No points, such as $x^2 + y^2 = -1$ – an imaginary ellipse.

Discriminant

Non-degenerate real conics can be classified as ellipses, parabolas, or hyperbolas by the discriminant of the non-homogeneous form $Ax^2 + 2Bxy + Cy^2 + 2Dx + 2Ey + F$, which is the determinant of the matrix

$$\begin{bmatrix} A & B \\ B & C \end{bmatrix},$$

the matrix of the quadratic form in (x, y).

Analogously, a conic can be classified as non-degenerate or degenerate according to the discriminant of the *homogeneous* quadratic form in (x, y, z). Here the affine form is homogenized to

$$Ax^2 + 2Bxy + Cy^2 + 2Dxz + 2Eyz + Fz^2;$$

the discriminant of this form is the determinant of the matrix:

$$\begin{bmatrix} A & B & D \\ B & C & E \\ D & E & F \end{bmatrix}.$$

The conic is degenerate if and only if the determinant of this matrix equals zero.

Relation to Intersection of a Plane and a Cone

Conics, also known as conic sections to emphasize their three-dimensional geometry, arise as the intersection of a plane with a cone. Degeneracy occurs when the plane contains the apex of the cone or when the cone degenerates to a cylinder and the plane is parallel to the axis of the cylinder.

Applications

Degenerate conics, as with degenerate algebraic varieties generally, arise as limits of non-degenerate conics, and are important in compactification of moduli spaces of curves.

For example, the pencil of curves (1-dimensional linear system of conics) defined by $x^2 + ay^2 = 1$ is non-degenerate for $a \neq 0$ but is degenerate for $a = 0$; concretely, it is an ellipse for $a > 0$, two parallel lines for $a = 0$, and a hyperbola with $a < 0$ – throughout, one axis has length 2 and the other has length $1/\sqrt{|a|}$, which is infinity for $a = 0$.

Such families arise naturally – given four points in general linear position (no three on a line), there is a pencil of conics through them (five points determine a conic, four points leave one parameter free), of which three are degenerate, each consisting of a pair of lines, corresponding to the $\binom{4}{2,2} = 3$ ways of choosing 2 pairs of points from 4 points (counting via the multinomial coefficient).

For example, given the four points $(\pm 1, \pm 1)$, the pencil of conics through them can be parameterized as $(1+a)x^2 + (1-a)y^2 = 2$, yielding the following pencil; in all cases the center is at the origin:

- $a > 1$: hyperbolae opening left and right;
- $a = 1$: the parallel vertical lines $x = -1, x = 1$;
- $0 < a < 1$: ellipses with a vertical major axis;

- $a = 0$: a circle (with radius $\sqrt{2}$);
- $-1 < a < 0$: ellipses with a horizontal major axis;
- $a = -1$: the parallel horizontal lines $y = -1, y = 1$;
- $a < -1$: hyperbolae opening up and down,
- $a = \infty$: the diagonal lines $y = x, y = -x$; (dividing by a and taking the limit as $a \to \infty$ yields $x^2 - y^2 = 0$)
- This then loops around to $a > 1$, since pencils are a *projective* line.

Note that this parametrization has a symmetry, where inverting the sign of a reverses x and y. In the terminology of (Levy 1964), this is a Type I linear system of conics, and is animated in the linked video.

A striking application of such a family is in (Faucette 1996) which gives a geometric solution to a quartic equation by considering the pencil of conics through the four roots of the quartic, and identifying the three degenerate conics with the three roots of the resolvent cubic.

Pappus's hexagon theorem is the special case of Pascal's theorem, when a conic degenerates to two lines.

Degeneration

In the complex projective plane, all conics are equivalent, and can degenerate to either two different lines or one double line.

In the real affine plane:

- Hyperbolae can degenerate to two intersecting lines (the asymptotes), as in $x^2 - y^2 = a^2$, or to two parallel lines: $x^2 - a^2y^2 = 1$, or to double line: $x^2 - a^2y^2 = a^2$,
- Parabolae can degenerate to two parallel lines: $x^2 - ay - 1 = 0$ or a double line $x^2 - ay = 0$, but, because parabolae have a double point at infinity, cannot degenerate to two intersecting lines.
- Ellipses can degenerate to two parallel lines: $x^2 + a^2y^2 - 1 = 0$ or a double line $x^2 + a^2y^2 - a^2 = 0$, but, because they have conjugate complex points at infinity which become a double point on degeneration, cannot degenerate to two intersecting lines. Complex plane do represent a conic section of imaginary fused state between an ellipse and a parabola.

Degenerate conics can degenerate further to more special degenerate conics, as indicated by the dimensions of the spaces and points at infinity.

- Two intersecting lines can degenerate to two parallel lines, by rotating until parallel, as in $x^2 - ay^2 - 1 = 0$, or to a double line by rotating into each other about a point, as in $x^2 - a^2 = 0$,
- Two parallel lines can degenerate to a double line by moving into each other, as in $x^2 - a^2 = 0$, but cannot degenerate to non-parallel lines.
- A double line cannot degenerate to the other types.
- Another type of degeneration occurs when an ellipse, rotated and translated to its simplest form $\frac{x^2}{a^2} + \frac{y^2}{b^2} = 1$, has its semiminor axis b go to zero and thus has its eccentricity go to one. The result is a line segment (degenerate because the ellipse is not differentiable at the endpoints) with its foci at the endpoints. As an orbit, this is a radial elliptic trajectory.

Points to Define

A general conic is defined by five points: given five points in general position, there is a unique conic passing through them. If three of these points lie on a line, then the conic is reducible, and may or may not be unique. If no four points are collinear, then five points define a unique conic (degenerate if three points are collinear, but the other two points determine the unique other line).

If four points are collinear, however, then there is not a unique conic passing through them – one line passing through the four points, and the remaining line passes through the other point, but the angle is undefined, leaving 1 parameter free. If all five points are collinear, then the remaining line is free, which leaves 2 parameters free.

Given four points in general linear position (no three collinear; in particular, no two coincident), there are exactly three pairs of lines (degenerate conics) passing through them, which will in general be intersecting, unless the points form a trapezoid (one pair is parallel) or a parallelogram (two pairs are parallel).

Given three points, if they are non-collinear, there are three pairs of parallel lines passing through them – choose two to define one line, and the third for the parallel line to pass through, by the parallel postulate.

Given two distinct points, there is a unique double line through them.

Eccentricity, Focus and Directrix

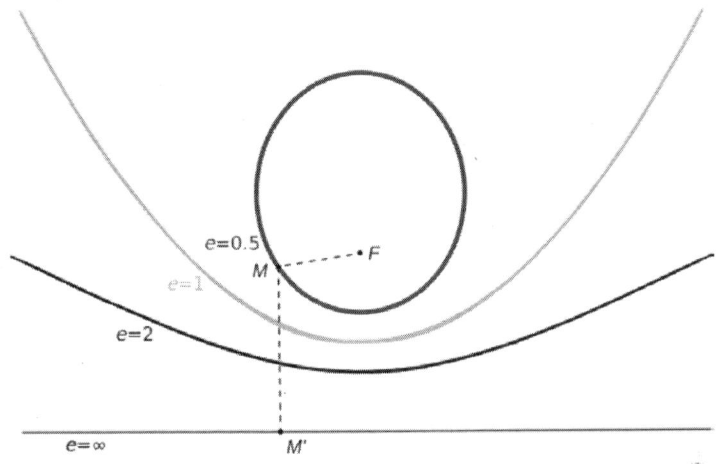

Figure: *Ellipse* (e=*1/2*)*, parabola* (e=*1*) *and hyperbola* (e=*2*) *with fixed focus* F *and directrix* (e=∞)*.*

The four defining conditions above can be combined into one condition that depends on a fixed point F (the *focus*), a line L (the *directrix*) not containing F and a nonnegative real number e (the eccentricity). The corresponding conic section consists of the locus of all points whose distance to F equals times their distance to L. For $0 < e < 1$ we obtain an ellipse, for $e = 1$ a parabola, and for $e > 1$ a hyperbola.

For an ellipse and a hyperbola, two focus-directrix combinations can be taken, each giving the same full ellipse or hyperbola. The distance from the center to the directrix is a / e, where is the semi-major axis of the ellipse, or the distance from the center to the tops of the hyperbola. The distance from the center to a focus is ae .

In the case of a circle, the eccentricity $e = 0$, and one can imagine the directrix to be infinitely far removed from the center. However, the statement that the circle consists of all points whose distance to F is e times the distance to L is not useful, because we get zero times infinity.

The eccentricity of a conic section is thus a measure of how far it deviates from being circular.

For a given a , the closer e is to 1, the smaller is the semi-minor axis.

Generalizations

Conics may be defined over other fields, and may also be classified in the projective plane rather than in the affine plane.

Over the complex numbers ellipses and hyperbolas are not distinct, since -1 is a square; precisely, the ellipse $x^2 + y^2 = 1$ becomes a hyperbola under the substitution geometrically a complex rotation, yielding $x^2 - w^2 = 1$ – a hyperbola is simply an ellipse with an imaginary axis length. Thus there is a 2-way classification: ellipse/hyperbola and parabola. Geometrically, this corresponds to intersecting the line at infinity in either 2 distinct points (corresponding to two asymptotes) or in 1 double point (corresponding to the axis of a parabola), and thus the real hyperbola is a more suggestive image for the complex ellipse/ hyperbola, as it also has 2 (real) intersections with the line at infinity.

In projective space, over any division ring, but in particular over either the real or complex numbers, all non-degenerate conics are equivalent, and thus in projective geometry one simply speaks of "a conic" without specifying a type, as type is not meaningful. Geometrically, the line at infinity is no longer special (distinguished), so while some conics intersect the line at infinity differently, this can be changed by a projective transformation – pulling an ellipse out to infinity or pushing a parabola off infinity to an ellipse or a hyperbola.

A generalization of a non degenerate conic in a projective plane is an oval. An oval is a point set that has the following properties, which are held by conics: 1) any line intersects an oval in none, one or two points, 2) at any point of the oval there exists a unique tangent line.

In Other Areas of Mathematics

The classification into elliptic, parabolic, and hyperbolic is pervasive in mathematics, and often divides a field into sharply distinct subfields. The classification mostly arises due to the presence of a quadratic form (in two variables this corresponds to the associated discriminant), but can also correspond to eccentricity.

Quadratic Form Classifications: quadratic forms

Quadratic forms over the reals are classified by Sylvester's law of inertia, namely by their positive index, zero index, and negative index: a quadratic form in n variables can be converted to a diagonal form, $x_1^2 + x_2^2 + \cdots + x_k^2 - x_{k+1}^2 - \cdots - x_{k+l}^2$, as where the number of +1 coefficients, *k*, is the positive index, the number of –1 coefficients, *l*, is the negative index, and the remaining variables are the zero index *m*, so $k + l + m = n$. In two variables the non-zero quadratic forms are classified as:

- $x^2 + y^2$, – positive-definite (the negative is also included), corresponding to ellipses,
- x^2 – degenerate, corresponding to parabolas, and
- $x^2 - y^2$ – indefinite, corresponding to hyperbolas.

In two variables quadratic forms are classified by discriminant, analogously to conics, but in higher dimensions the more useful classification is as *definite*, (all positive or all negative), *degenerate*, (some zeros), or *indefinite* (mix of positive and negative but no zeros). This classification underlies many that follow.

Curvature

The Gaussian curvature of a surface describes the infinitesimal geometry, and may at each point be either positive – elliptic geometry, zero – Euclidean geometry (flat, parabola), or negative – hyperbolic geometry; infinitesimally, to second order the surface looks like the graph of $x^2 + y^2$,, x^2 (or 0), or $x^2 - y^2$. Indeed, by the uniformization theorem every surface can be taken to be globally (at every point) positively curved, flat, or negatively curved. In higher dimensions the Riemann curvature tensor is a more complicated object, but manifolds with constant sectional curvature are interesting objects of study, and have strikingly different properties, as discussed at sectional curvature.

Second Order PDEs

Partial differential equations (PDEs) of second order are classified at each point as elliptic, parabolic, or hyperbolic, accordingly as their second order terms correspond to an elliptic, parabolic, or hyperbolic quadratic form. The behaviour and theory of these different types of PDEs are strikingly different – representative examples is that the Poisson equation is elliptic, the heat equation is parabolic, and the wave equation is hyperbolic.

Eccentricity Classifications Include: Möbius transformations

Real Möbius transformations (elements of PSL2(R) or its 2-fold cover, SL2(R)) are classified as elliptic, parabolic, or hyperbolic accordingly as their half-trace is $0 \leq |\mathrm{tr}|/2 < 1$, $|\mathrm{tr}|/2 = 1$, or $|\mathrm{tr}|/2 > 1$, mirroring the classification by eccentricity.

Variance-to-mean Ratio

The variance-to-mean ratio classifies several important families of discrete probability distributions: the constant distribution as circular (eccentricity 0), binomial distributions as elliptical, Poisson distributions as parabolic, and negative binomial distributions as hyperbolic. This is elaborated at cumulants of some discrete probability distributions.

Cartesian Coordinates

In the Cartesian coordinate system, the graph of a quadratic equation in two variables is always a conic section – though it may be degenerate, and all conic sections arise in this way. The equation will be of the form

$$Ax^2 + Bxy + Cy^2 + Dx + Ey + F = 0 \text{ with } A, B, C \text{ not all zero.}$$

As scaling all six constants yields the same locus of zeros, one can consider conics as points in the five-dimensional projective space $\mathbf{P}^5$.

Discriminant Classification

The conic sections described by this equation can be classified with the discriminant

$$B^2 - 4AC.$$

If the conic is non-degenerate, then:

- if $B^2 - 4AC < 0$, the equation represents an ellipse;
  - if $A = C$ and $B = 0$, the equation represents a circle, which is a special case of an ellipse;
- if $B^2 - 4AC = 0$, the equation represents a parabola;
- if $B^2 - 4AC > 0$, the equation represents a hyperbola;
  - if we also have $A + C = 0$, the equation represents a rectangular hyperbola.

To distinguish the degenerate cases from the non-degenerate cases, let Δ be the determinant of the 3×3 matrix $[A, B/2, D/2 ; B/2, C, E/2 ; D/2, E/2, F]$: that is, $\Delta = (AC - B^2/4)F + BED/4 - CD^2/4 - AE^2/4$. Then the conic section is non-degenerate if and only if $\Delta \neq 0$. If $\Delta=0$ we have a point ellipse, two parallel lines (possibly coinciding with each other) in the case of a parabola, or two intersecting lines in the case of a hyperbola.

Moreover, in the case of a non-degenerate ellipse (with $B^2 - 4AC < 0$ and $\Delta\neq0$), we have a real ellipse if $C\Delta < 0$ but an imaginary ellipse if $C\Delta > 0$. An example is , which has no real-valued solutions.

Note that A and B are polynomial coefficients, not the lengths of semi-major/minor axis as defined in some sources.

Matrix Notation

The above equation can be written in matrix notation as

$$\begin{bmatrix} x & y \end{bmatrix} \cdot \begin{bmatrix} A & B/2 \\ B/2 & C \end{bmatrix} \cdot \begin{bmatrix} x \\ y \end{bmatrix} + Dx + Ey + F = 0.$$

The type of conic section is solely determined by the determinant of middle matrix: if it is positive, zero, or negative then the conic is an ellipse, parabola, or hyperbola respectively. If both the eigenvalues of the middle matrix are non-zero (i.e. it is an ellipse or a hyperbola), we can do a transformation of variables to obtain

$$\begin{pmatrix} x-a \\ y-c \end{pmatrix}^T \begin{pmatrix} A & \frac{B}{2} \\ \frac{B}{2} & C \end{pmatrix} \begin{pmatrix} x-a \\ y-c \end{pmatrix} = G$$

where a,c, and G satisfy $D + 2aA + Bc = 0, E + 2Cc + Ba = 0$, and .

$$G = Aa^2 + Cc^2 + Bac - F$$

The quadratic can also be written as

$$\begin{bmatrix} x & y & 1 \end{bmatrix} \cdot \begin{bmatrix} A & B/2 & D/2 \\ B/2 & C & E/2 \\ D/2 & E/2 & F \end{bmatrix} \cdot \begin{bmatrix} x \\ y \\ 1 \end{bmatrix} = 0.$$

If the determinant of this 3×3 matrix is non-zero, the conic section is not degenerate. If the determinant equals zero, the conic is a degenerate parabola (two parallel or coinciding lines), a degenerate ellipse (a point ellipse), or a degenerate hyperbola (two intersecting lines).

Note that in the centered equation with constant term G, G equals minus one times the ratio of the 3×3 determinant to the 2×2 determinant.

As Slice of Quadratic Form

The equation

$$Ax^2 + Bxy + Cy^2 + Dx + Ey + F = 0$$

can be rearranged by taking the affine linear part to the other side, yielding

$$Ax^2 + Bxy + Cy^2 = -(Dx + Ey + F).$$

In this form, a conic section is realised exactly as the intersection of the graph of the quadratic form $z = Ax^2 + Bxy + Cy^2$ and the plane $z = -(Dx + Ey + F)$. Parabolas and hyperbolas can be realised by a horizontal plane ($D = E = 0$), while ellipses require that the plane be slanted. Degenerate conics correspond to degenerate intersections, such as taking slices such as $z = -1$ of a positive-definite form.

Eccentricity in Terms of Parameters of the Quadratic Form

When the conic section is written algebraically as

$$Ax^2 + Bxy + Cy^2 + Dx + Ey + F = 0,$$

the eccentricity can be written as a function of the parameters of the quadratic equation. If $4AC = B^2$ the conic is a parabola and its eccentricity equals 1 (if it is non-degenerate). Otherwise, assuming the equation represents either a non-degenerate hyperbola or a non-degenerate, non-imaginary ellipse, the eccentricity is given by

$$e = \sqrt{\frac{2\sqrt{(A-C)^2 + B^2}}{\eta(A+C) + \sqrt{(A-C)^2 + B^2}}},$$

where $\eta = 1$ if the determinant of the 3×3 matrix is negative and $\eta = -1$ if that determinant is positive.

Standard Form

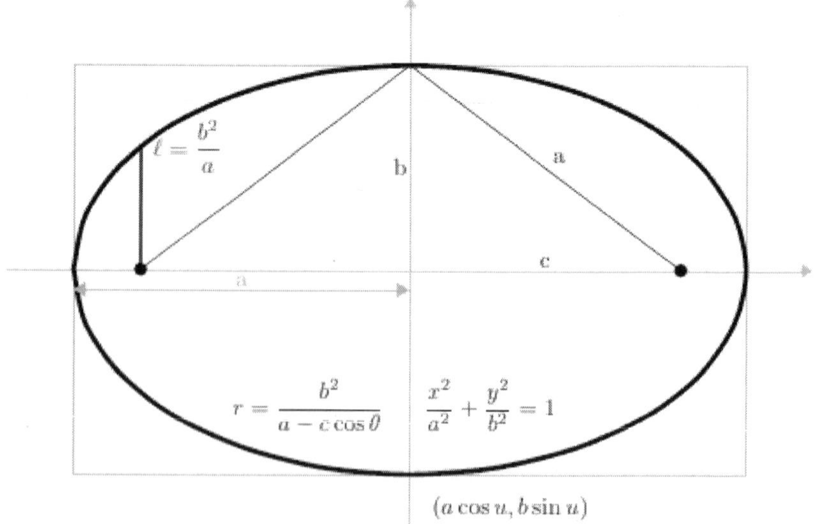

Figure: *Standard forms of an ellipse*

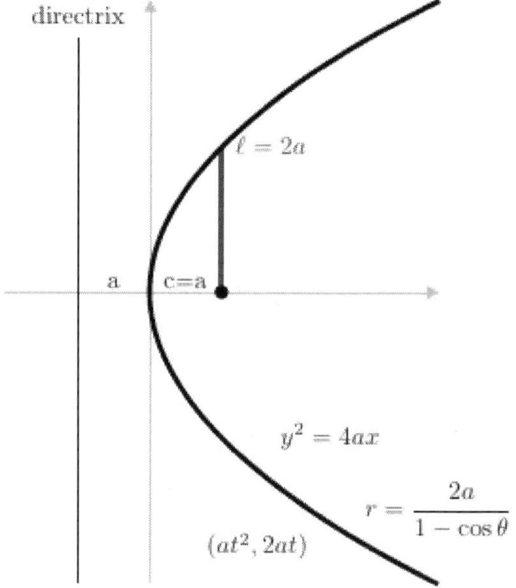

Figure: *Standard forms of a parabola*

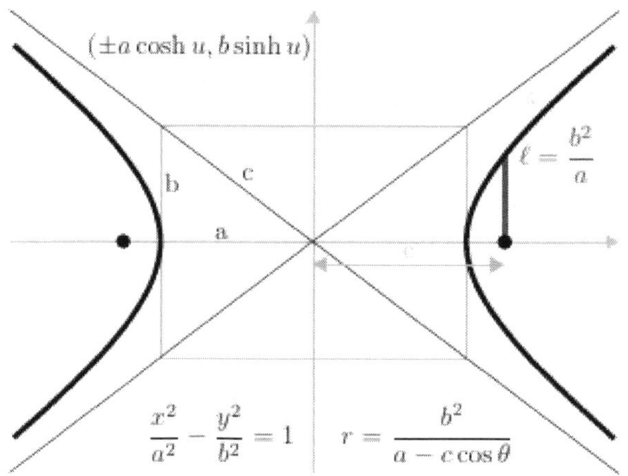

Figure: *Standard forms of a hyperbola*

Through change of coordinates (a rotation of axes and a translation) these equations can be put in standard forms:

- Circle: $x^2 + y^2 = a^2$
- Ellipse: $x^2/a^2 + y^2/b^2 = 1$
- Parabola: $y^2 = 4ax$, $x^2 = 4ay$
- Hyperbola: $x^2/a^2 - y^2/b^2 = 1$, $x^2/b^2 - y^2/a^2 = -1$
- Rectangular hyperbola: $xy = c^2$

Such forms will be symmetrical about both the x-axis and y-axis (for the circle, ellipse and hyperbola), or about either but not both (for the parabola). The rectangular hyperbola, however, is only symmetrical about the lines $y = x$ and $y = -x$.

These standard forms can be written as parametric equations,

- Circle: $(a \cos \theta, a \sin \theta)$,
- Ellipse: $(a \cos \theta, b \sin \theta)$,
- Parabola: $(at^2, 2at)$,
- Hyperbola: $(a \sec \theta, b \tan \theta)$ or $(\pm a \cosh u, b \sinh u)$,
- Rectangular hyperbola: $(ct, c/t)$.

Invariants of Conics

The trace and determinant of $\begin{bmatrix} A & B/2 \\ B/2 & C \end{bmatrix}$ are both invariant with respect to both rotation of axes and translation of the plane (movement of the origin). The constant term F is invariant under rotation only.

Modified Form

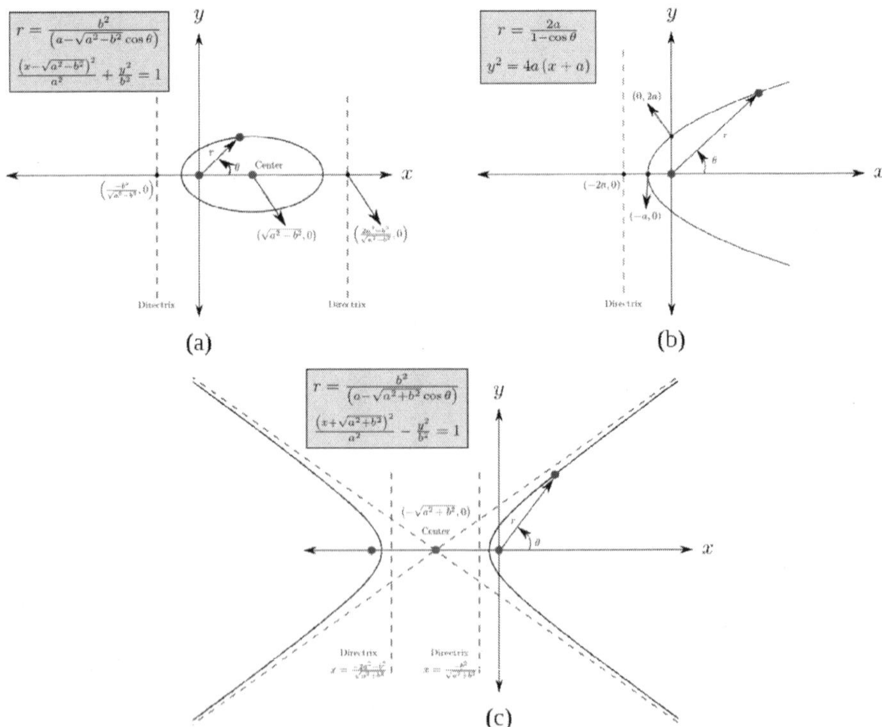

Figure: *Three different types of conic sections. Focal-points corresponding to all conic sections are placed at the origin.*

For some practical applications, it is important to re-arrange the standard form so that the focal-point can be placed at the origin. The mathematical formulation for a general conic section is then given in the polar form by

$$r = \frac{l}{1 - e\cos\theta}$$

and in the Cartesian form by

$$\sqrt{x^2 + y^2} = (l + ex)$$

$$\Rightarrow \left(\frac{x - \dfrac{le}{1-e^2}}{\dfrac{l}{1-e^2}} \right)^2 + \frac{(1-e^2)y^2}{l^2} = 1$$

From the above equation, the linear eccentricity (c) is given by $c = \left(\frac{le}{1-e^2} \right)$.

From the general equations given above, different conic sections can be represented as shown below:

- Circle: $x^2 + y^2 = r^2$
- Ellipse: $\dfrac{\left(x - \sqrt{a^2 - b^2}\right)^2}{a^2} + \dfrac{y^2}{b^2} = 1$
- Parabola: $y^2 = 4a(x + a)$
- Hyperbola: $\dfrac{\left(x + \sqrt{a^2 + b^2}\right)^2}{a^2} - \dfrac{y^2}{b^2} = 1$

Homogeneous Coordinates

In homogeneous coordinates a conic section can be represented as:

$$A_1x^2 + A_2y^2 + A_3z^2 + 2B_1xy + 2B_2xz + 2B_3yz = 0.$$

Or in matrix notation

$$[x \quad y \quad z] \cdot \begin{bmatrix} A_1 & B_1 & B_2 \\ B_1 & A_2 & B_3 \\ B_2 & B_3 & A_3 \end{bmatrix} \cdot \begin{bmatrix} x \\ y \\ z \end{bmatrix} = 0.$$

The matrix $M = \begin{bmatrix} A_1 & B_1 & B_2 \\ B_1 & A_2 & B_3 \\ B_2 & B_3 & A_3 \end{bmatrix}$ is called *the matrix of the conic section.*

$\Delta = \det(M) = \det\left(\begin{bmatrix} A_1 & B_1 & B_2 \\ B_1 & A_2 & B_3 \\ B_2 & B_3 & A_3 \end{bmatrix}\right)$ is called the determinant of the conic section. If $\Delta = 0$ then the *conic section* is said to be *degenerate*; this means that the conic section is either a union of two straight lines, a repeated line, a point or the empty set.

For example, the conic section $\begin{bmatrix} x & y & z \end{bmatrix}.\begin{bmatrix} 1 & 0 & 0 \\ 0 & -1 & 0 \\ 0 & 0 & 0 \end{bmatrix}.\begin{bmatrix} x \\ y \\ z \end{bmatrix} = 0$ reduces to the union of two lines:

$$\{x^2 - y^2 = 0\} = \{(x+y)(x-y) = 0\} = \{x+y=0\} \cup \{x-y=0\}.$$

Similarly, a conic section sometimes reduces to a (single) repeated line:

$$\{x^2 + 2xy + y^2 = 0\} = \{(x+y)^2 = 0\} = \{x+y=0\} \cup \{x+y=0\} = \{x+y=0\}.$$

$\delta = \det\left(\begin{bmatrix} A_1 & B_1 \\ B_1 & A_2 \end{bmatrix}\right)$ is called the discriminant of the conic section. If $\delta = 0$ then the *conic section* is a parabola, if $\delta < 0$, it is an hyperbola and if $\delta > 0$, it is an ellipse. A conic section is a circle if $\delta > 0$ and $A_1 = A_2$ and $B_1 = 0$, it is an rectangular hyperbola if $\delta < 0$ and $A_1 = -A_2$. It can be proven that in the complex projective plane CP^2 two conic sections have four points in common (if one accounts for multiplicity), so there are never more than 4 intersection points and there is always one *intersection point* (possibilities: four distinct intersection points, two singular intersection points and one double intersection points, two double intersection points, one singular intersection point and 1 with multiplicity 3, 1 intersection point with multiplicity 4). If there exists at least one intersection point with multiplicity > 1, then the two conic sections are said to be tangent. If there is only one intersection point, which has multiplicity 4, the two conic sections are said to be osculating.

Furthermore each straight line intersects each conic section twice. If the intersection point is double, the line is said to be tangent and it is called the tangent line. Because every straight line intersects a conic section twice, each conic section has two points at infinity (the intersection points with the line at infinity). If these points are real, the conic section must be a hyperbola, if they are imaginary conjugated, the conic section must be an ellipse, if the conic section has one double point at infinity it is a parabola. If the points at infinity are (1,i,0) and (1,-i,0), the conic section is a circle. If a conic section has one real and one imaginary point at infinity or it has two imaginary points that are not conjugated then it not a real conic section (its coefficients are complex).

Degenerate Cases

There are five degenerate cases: three in which the plane passes through apex of the cone, and three that arise when the cone itself degenerates to a cylinder (a doubled line can occur in both cases). When the plane passes through the apex, the resulting conic is always degenerate, and is either: a point (when the angle between the plane and the axis of the cone is larger than tangential); a straight line (when the plane is tangential to the surface of the cone); or a pair of intersecting lines (when the angle is smaller than the tangential). These correspond respectively to degeneration of an ellipse, parabola, and a hyperbola, which are characterized in the same way by angle. The straight line is more precisely a *double* line (a line with multiplicity 2) because the plane is tangent to the cone, and thus the intersection should be counted twice.

Where the cone is a cylinder, i.e. with the vertex at infinity, cylindric sections are obtained; this corresponds to the apex being at infinity. Cylindrical sections are ellipses (or circles), unless the plane is vertical (which corresponds to passing through the apex at infinity), in which case three degenerate cases occur: two parallel lines, known as a ribbon (corresponding to an ellipse with one axis infinite and the other axis real and non-zero, the distance between the lines), a double line (an ellipse with one infinite axis and one axis zero), and no intersection (an ellipse with one infinite axis and the other axis imaginary).

Three Dimension, an Introduction

Three-dimensional space is a geometric three-parameter model of the physical universe (without considering time) in which all known matter exists. These three dimensions can be labelled by a combination of three chosen from the terms length, width, height, depth, and breadth. Any three directions can be chosen, provided that they do not all lie in the same plane.

In physics and mathematics, a sequence of n numbers can be understood as a location in n-dimensional space. When $n = 3$, the set of all such locations is called three-dimensional Euclidean space. It is commonly represented by the symbol $\mathbb{R}^3$. This space is only one example of a great variety of spaces in three dimensions called 3-manifolds.

In mathematics, analytic geometry (also called Cartesian geometry) describes any point in three-dimensional space by means of three coordinates. Three coordinate axes are given, usually each perpendicular to the other two at the origin, the point at which they cross. They are usually labelled x, y, and z. Relative to these axes, the position of any point in three-dimensional space is given by an ordered triple of real numbers, each number giving the distance of that point from the origin measured along the given axis, which is equal to the distance of that point from the plane determined by the other two axes.

Coordinate Systems

In mathematics, analytic geometry (also called Cartesian geometry) describes every point in three-dimensional space by means of three coordinates. Three coordinate axes are given, each perpendicular to the other two at the origin, the point at which they cross. They are usually labelled x, y, and z. Relative to these axes, the position of any point in three-dimensional space is given by an ordered triple of real numbers, each number giving the distance of that point from the origin measured along the given axis, which is equal to the distance of that point from the plane determined by the other two 2 axes.

Other popular methods of describing the location of a point in three-dimensional space include cylindrical coordinates and spherical coordinates, though there is an infinite number of possible methods.

Below are Images of the Above-Mentioned Systems.

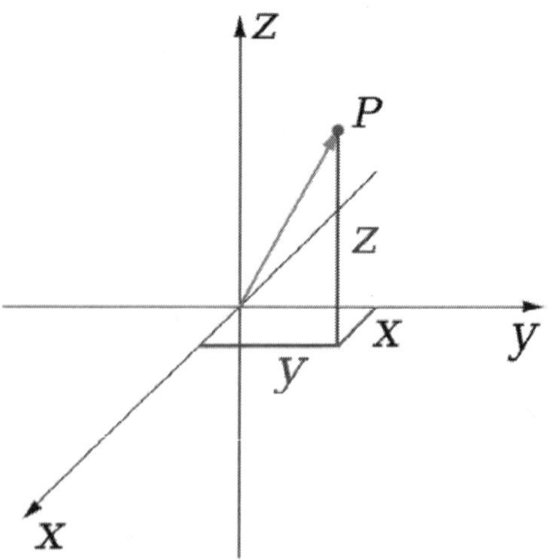

Figure: *Cartesian coordinate system*

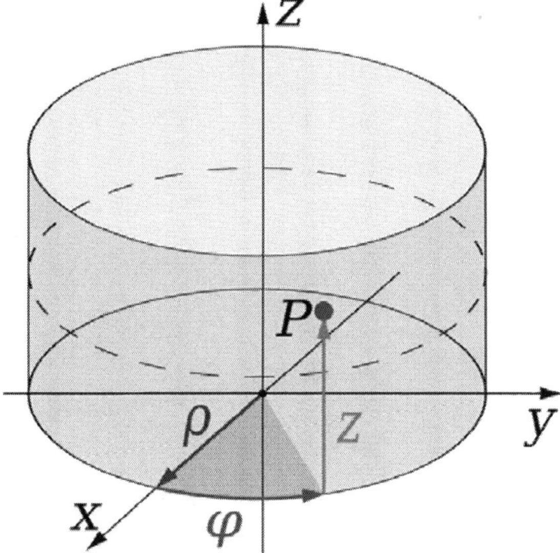

Figure: *Cylindrical coordinate system*

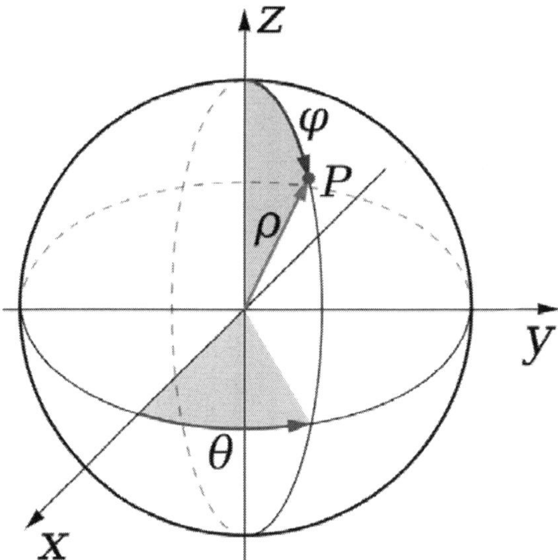

Figure: *Spherical coordinate system*

Polar Coordinates

The polar coordinate system is a two-dimensional coordinate system in which each point on a plane is determined by a distance from a fixed point and an angle from a fixed direction.

The fixed point (analogous to the origin of a Cartesian system) is called the *pole*, and the ray from the pole in the fixed direction is the *polar axis*. The distance from the pole is called the *radial coordinate* or *radius*, and the angle is the *angular coordinate, polar angle,* or azimuth.

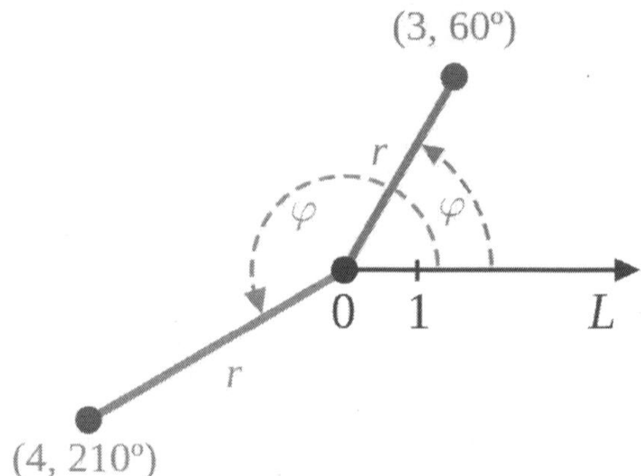

Figure: *Points in the polar coordinate system with pole O and polar axis L. In green, the point with radial coordinate 3 and angular coordinate 60 degrees, or (3,60°). In blue, the point (4,210°).*

The concepts of angle and radius were already used by ancient peoples of the 1st millennium BC. The Greek astronomer and astrologer Hipparchus (190–120 BC) created a table of chord functions giving the length of the chord for each angle, and there are references to his using polar coordinates in establishing stellar positions. In On Spirals, Archimedes describes the Archimedean spiral, a function whose radius depends on the angle. The Greek work, however, did not extend to a full coordinate system.

From the 8th century AD onward, astronomers developed methods for approximating and calculating the direction to Makkah (qibla)—and its distance—from any location on the Earth. From the 9th century onward they were using spherical trigonometry and map projection methods to determine these quantities accurately. The calculation is essentially the conversion of the equatorial polar coordinates of Mecca (i.e. its longitude and latitude) to its polar coordinates (i.e. its qibla and distance) relative to a system whose reference meridian is the great circle through the given location and the Earth's poles, and whose polar axis is the line through the location and its antipodal point.

There are various accounts of the introduction of polar coordinates as part of a formal coordinate system. The full history of the subject is described in Harvard professor Julian Lowell Coolidge's *Origin of Polar Coordinates.* Grégoire de Saint-Vincent and Bonaventura Cavalieri independently introduced the concepts in the mid-seventeenth century. Saint-Vincent wrote about them privately in 1625 and published his work in 1647, while Cavalieri published his in 1635 with a corrected version appearing in 1653. Cavalieri first used polar coordinates to solve a problem relating to the area within an Archimedean spiral. Blaise Pascal subsequently used polar coordinates to calculate the length of parabolic arcs.

In Method of Fluxions (written 1671, published 1736), Sir Isaac Newton examined the transformations between polar coordinates, which he referred to as the "Seventh Manner; For Spirals", and nine other coordinate systems. In the journal Acta Eruditorum (1691), Jacob Bernoulli used a system with a point on a line, called the *pole* and *polar axis* respectively. Coordinates were specified by the distance from the pole and the angle from the *polar axis*. Bernoulli's work extended to finding the radius of curvature of curves expressed in these coordinates.

The actual term *polar coordinates* has been attributed to Gregorio Fontana and was used by 18th-century Italian writers. The term appeared in English in George Peacock's 1816 translation of Lacroix's *Differential and Integral Calculus.* Alexis Clairaut was the first to think of polar coordinates in three dimensions, and Leonhard Euler was the first to actually develop them.

Conventions

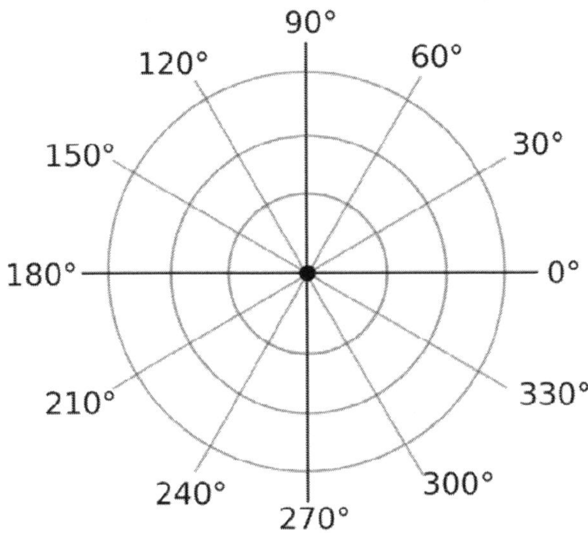

***Figure:** A polar grid with several angles labelled in degrees*

The radial coordinate is often denoted by *r*, and the angular coordinate by φ, θ, or *t*. The angular coordinate is specified as *ö* by ISO standard 31-11.

Angles in polar notation are generally expressed in either degrees or radians (2π rad being equal to 360°). Degrees are traditionally used in navigation, surveying, and many applied disciplines, while radians are more common in mathematics and mathematical physics.

In many contexts, a positive angular coordinate means that the angle φ is measured counterclockwise from the axis.

In mathematical literature, the polar axis is often drawn horizontal and pointing to the right.

Uniqueness of Polar Coordinates

Adding any number of full turns (360°) to the angular coordinate does not change the corresponding direction. Also, a negative radial coordinate is best interpreted as the corresponding positive distance measured in the opposite direction. Therefore, the same point can be expressed with an infinite number of different polar coordinates $(r, \varphi \pm n\times 360°)$ or $(-r, \varphi \pm (2n + 1)180°)$, where n is any integer. Moreover, the pole itself can be expressed as $(0, \varphi)$ for any angle φ.

Where a unique representation is needed for any point, it is usual to limit r to non-negative numbers ($r \geq 0$) and φ to the interval [0, 360°) or (–180°, 180°] (in radians, $[0, 2\pi)$ or $(-\pi, \pi]$). One must also choose a unique azimuth for the pole, e.g., $\varphi = 0$.

Converting between Polar and Cartesian Coordinates

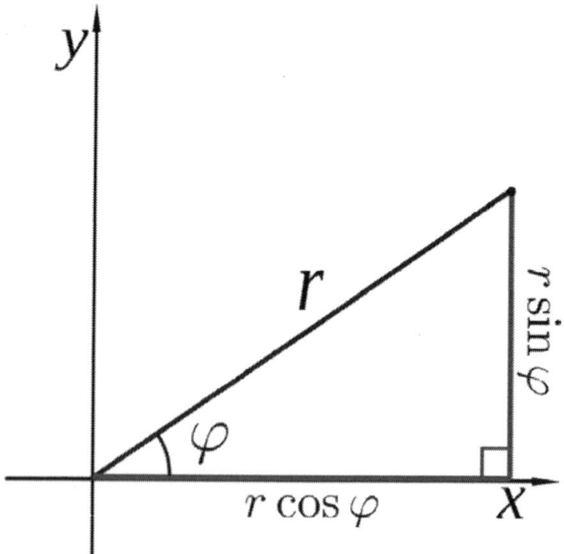

Figure: *A diagram illustrating the relationship between polar and Cartesian coordinates.*

The polar coordinates r and φ can be converted to the Cartesian coordinates x and y by using the trigonometric functions sine and cosine:

$$x = r\cos\varphi$$

$$y = r\sin\varphi$$

The Cartesian coordinates x and y can be converted to polar coordinates r and ö with $r \geq 0$ and φ in the interval $(-\pi, \pi]$ by:

$r = \sqrt{x^2 + y^2}$ (as in the Pythagorean theorem or the Euclidean norm), and

$$\varphi = \operatorname{atan} 2(y, x) \quad ,$$

where atan2 is a common variation on the arctangent function defined as

$$\operatorname{atan} 2(y,x) = \begin{cases} \arctan(\frac{y}{x}) & \text{if } x > 0 \\ \arctan(\frac{y}{x}) + \pi & \text{if } x < 0 \text{ and } y \geq 0 \\ \arctan(\frac{y}{x}) - \pi & \text{if } x < 0 \text{ and } y < 0 \\ \frac{\pi}{2} & \text{if } x = 0 \text{ and } y > 0 \\ -\frac{\pi}{2} & \text{if } x = 0 \text{ and } y < 0 \\ \text{undefined} & \text{if } x = 0 \text{ and } y = 0 \end{cases}$$

The value of φ above is the principal value of the complex number function arg applied to $x+iy$. An angle in the range $[0, 2\pi)$ may be obtained by adding 2π to the value in case it is negative.

Polar Equation of a Curve

The equation defining an algebraic curve expressed in polar coordinates is known as a *polar equation*. In many cases, such an equation can simply be specified by defining r as a function of φ. The resulting curve then consists of points of the form $(r(\varphi), \varphi)$ and can be regarded as the graph of the polar function r.

Different forms of symmetry can be deduced from the equation of a polar function r. If $r(-\varphi) = r(\varphi)$ the curve will be symmetrical about the horizontal (0°/180°) ray, if $r(\pi - \varphi) = r(\varphi)$ it will be symmetric about the vertical (90°/270°) ray, and if $r(\varphi - \alpha) = r(\varphi)$ it will be rotationally symmetric by α counterclockwise about the pole.

Because of the circular nature of the polar coordinate system, many curves can be described by a rather simple polar equation, whereas their Cartesian form is much more intricate. Among the best known of these curves are the polar rose, Archimedean spiral, lemniscate, limaηon, and cardioid.

For the circle, line, and polar rose below, it is understood that there are no restrictions on the domain and range of the curve.

Circle

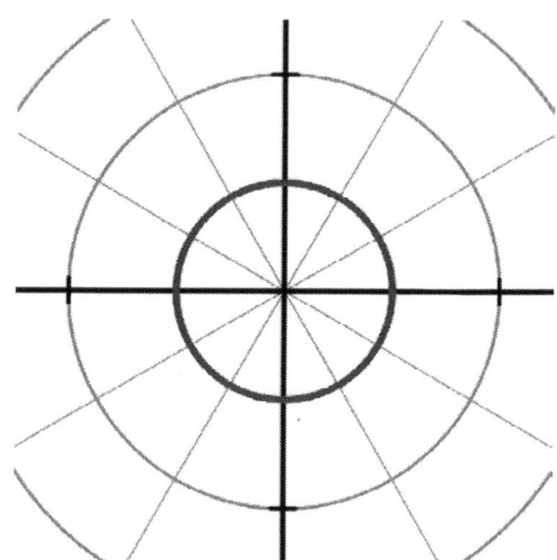

Figure: *A circle with equation* $r(\varphi) = 1$

The general equation for a circle with a center at (r_0, γ) and radius a is

$$r^2 - 2rr_0\cos(\varphi - \gamma) + r_0^2 = a^2.$$

This can be simplified in various ways, to conform to more specific cases, such as the equation

$$r(\varphi) = a$$

for a circle with a center at the pole and radius a.

When r_0 = a, or when the origin lies on the circle, the equation becomes

$$r = 2a\cos(\varphi - \gamma).$$

In the general case, the equation can be solved for r, giving

$$r = r_0\cos(\varphi - \gamma) + \sqrt{a^2 - r_0^2\sin^2(\varphi - \gamma)},$$

the solution with a minus sign in front of the square root gives the same curve.

Line

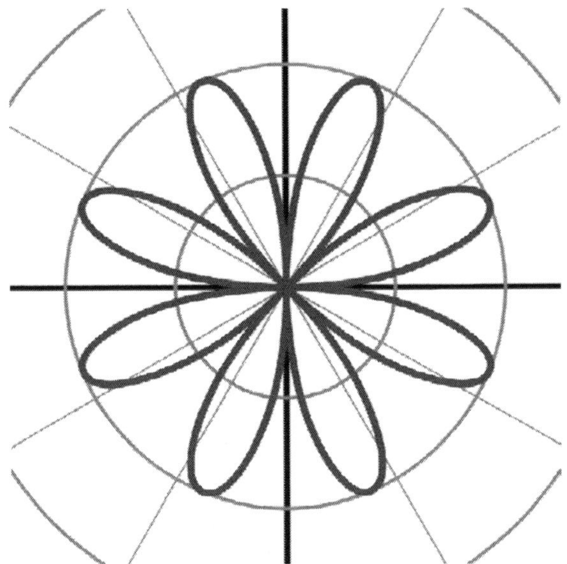

Figure: *A polar rose withs equation* r(φ) = *2 sin 4*φ

Radial lines (those running through the pole) are represented by the equation

$$\varphi = \gamma$$

where c is the angle of elevation of the line; that is, c = arctan *m* where *m* is the slope of the line in the Cartesian coordinate system. The non-radial line that crosses the radial line φ = c perpendicularly at the point (r_0, c) has the equation

$$r(\varphi) = r_0 \sec(\varphi - \gamma).$$

Otherwise stated (r_0, c) is the point in which the tangent intersects the imaginary circle of radius r_0.

Polar Rose

A polar rose is a famous mathematical curve that looks like a petaled flower, and that can be expressed as a simple polar equation,

$$r(\varphi) = a\cos(k\varphi + \gamma_0)$$

for any constant c_0 (including 0). If *k* is an integer, these equations will produce a *k*-petaled rose if *k* is odd, or a 2*k*-petaled rose if *k* is even. If *k* is rational but not an integer, a rose-like shape may form but with overlapping petals. Note that these equations never define a rose with 2, 6, 10, 14, etc. petals. The variable *a* represents the length of the petals of the rose.

Archimedean Spiral

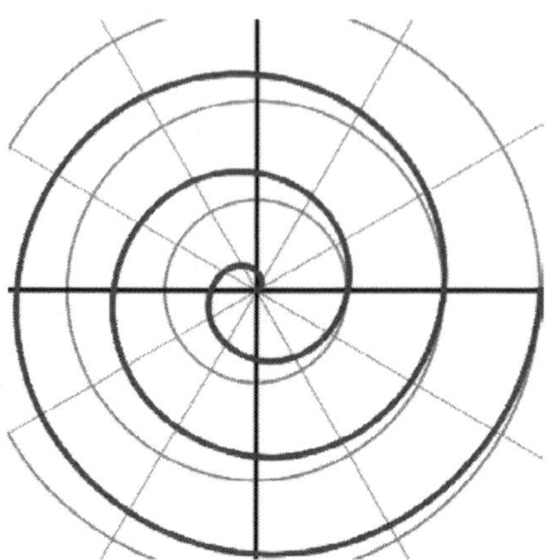

One arm of an Archimedean spiral with equation $r(\varphi) = \varphi / 2\pi$ for $0 < \varphi < 6\pi$

The Archimedean spiral is a famous spiral that was discovered by Archimedes, which can also be expressed as a simple polar equation. It is represented by the equation

$$r(\varphi) = a + b\varphi.$$

Changing the parameter a will turn the spiral, while b controls the distance between the arms, which for a given spiral is always constant. The Archimedean spiral has two arms, one for $\varphi > 0$ and one for $\varphi < 0$. The two arms are smoothly connected at the pole. Taking the mirror image of one arm across the 90°/270° line will yield the other arm. This curve is notable as one of the first curves, after the conic sections, to be described in a mathematical treatise, and as being a prime example of a curve that is best defined by a polar equation.

Conic Sections

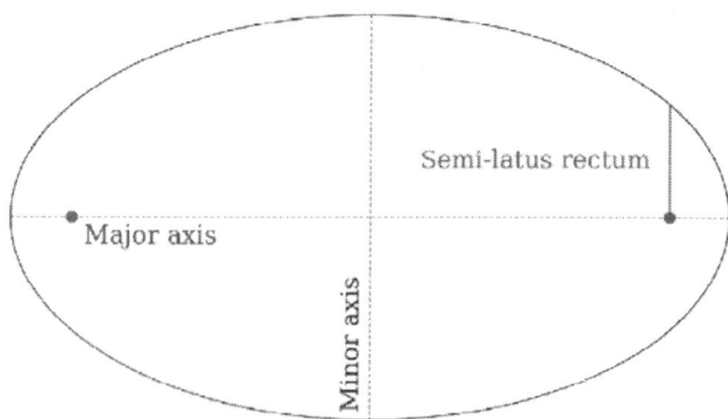

Figure: *Ellipse, showing semi-latus rectum*

A conic section with one focus on the pole and the other somewhere on the 0° ray (so that the conic's major axis lies along the polar axis) is given by:

$$r = \frac{\ell}{1 - e\cos\varphi}$$

where e is the eccentricity and ℓ is the semi-latus rectum (the perpendicular distance at a focus from the major axis to the curve). If $e > 1$, this equation defines a hyperbola; if $e = 1$, it defines a parabola; and if $e < 1$, it defines an ellipse. The special case $e = 0$ of the latter results in a circle of radius .

Intersection of Two Polar Curves

The graphs of two polar functions $r = f(\theta)$ and $r = g(\theta)$ have possible intersections in 3 cases:

1. In the origin if the equations $f(\theta) = 0$ and $g(\theta) = 0$ have at least one solution each.
2. All the points $[g(\theta_i), \theta_i]$ where θ_i are the solutions to the equation .
3. All the points $[g(\theta_i), \theta_i]$ where θ_i are the solutions to the equation $f(\theta + (2k+1)\pi) = -g(\theta)$ where is an integer.

Complex Numbers

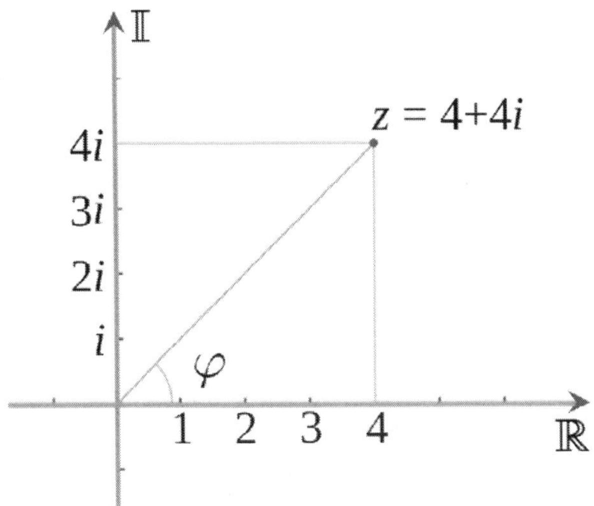

Figure: *An illustration of a complex number z plotted on the complex plane*

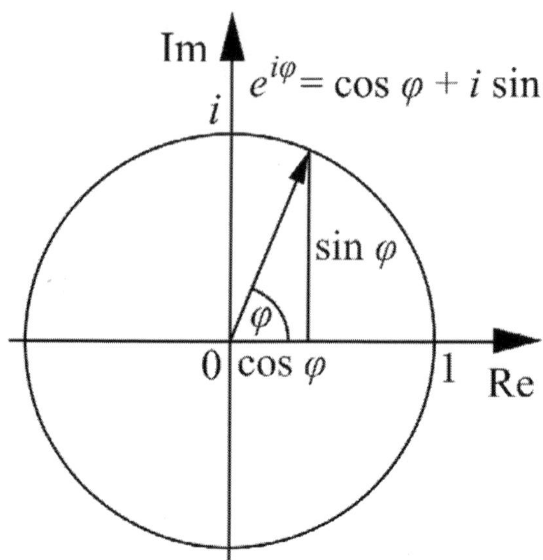

Figure: *An illustration of a complex number plotted on the complex plane using Euler's formula*

Every complex number can be represented as a point in the complex plane, and can therefore be expressed by specifying either the point's Cartesian coordinates (called rectangular or Cartesian form) or the point's polar coordinates (called polar form). The complex number z can be represented in rectangular form as

$$z = x + iy$$

where i is the imaginary unit, or can alternatively be written in polar form (via the conversion formulae given above) as

$$z = r \cdot (\cos\varphi + i\sin\varphi)$$

and from there as

$$z = re^{i\varphi}$$

where e is Euler's number, which are equivalent as shown by Euler's formula. (Note that this formula, like all those involving exponentials of angles, assumes that the angle φ is expressed in radians.) To convert between the rectangular and polar forms of a complex number, the conversion formulae given above can be used.

For the operations of multiplication, division, and exponentiation of complex numbers, it is generally much simpler to work with complex numbers expressed in polar form rather than rectangular form. From the laws of exponentiation:

- Multiplication: $r_0 e^{i\varphi_0} \cdot r_1 e^{i\varphi_1} = r_0 r_1 e^{i(\varphi_0 + \varphi_1)}$

- Division: $\dfrac{r_0 e^{i\varphi_0}}{r_1 e^{i\varphi_1}} = \dfrac{r_0}{r_1} e^{i(\varphi_0 - \varphi_1)}$
- Exponentiation (De Moivre's formula): $(re^{i\varphi})^n = r^n e^{in\varphi}$

Calculus

Calculus can be applied to equations expressed in polar coordinates.

The angular coordinate φ is expressed in radians throughout this section, which is the conventional choice when doing calculus.

Differential Calculus

Using $x = r\cos\varphi$ and $y = r\sin\varphi$, one can derive a relationship between derivatives in Cartesian and polar coordinates. For a given function, $u(x,y)$, it follows that (by computing its total derivatives)

$$r\frac{\partial u}{\partial r} = r\frac{\partial u}{\partial x}\frac{\partial x}{\partial r} + r\frac{\partial u}{\partial y}\frac{\partial y}{\partial r},$$

$$\frac{\partial u}{\partial \varphi} = \frac{\partial u}{\partial x}\frac{\partial x}{\partial \varphi} + \frac{\partial u}{\partial y}\frac{\partial y}{\partial \varphi},$$

or

$$r\frac{\partial u}{\partial r} = r\frac{\partial u}{\partial x}\cos\varphi + r\frac{\partial u}{\partial y}\sin\varphi = x\frac{\partial u}{\partial x} + y\frac{\partial u}{\partial y},$$

$$\frac{\partial u}{\partial \varphi} = -\frac{\partial u}{\partial x}r\sin\varphi + \frac{\partial u}{\partial y}r\cos\varphi = -y\frac{\partial u}{\partial x} + x\frac{\partial u}{\partial y}.$$

Hence, we have the following formulae:

$$r\frac{\partial}{\partial r} = x\frac{\partial}{\partial x} + y\frac{\partial}{\partial y}$$

$$\frac{\partial}{\partial \varphi} = -y\frac{\partial}{\partial x} + x\frac{\partial}{\partial y}.$$

Using the inverse coordinates transformation, an analogous reciprocal relationship can be derived between the derivatives. Given a function $u(r,\varphi)$, it follows that

$$\frac{\partial u}{\partial x} = \frac{\partial u}{\partial r}\frac{\partial r}{\partial x} + \frac{\partial u}{\partial \varphi}\frac{\partial \varphi}{\partial x},$$

$$\frac{\partial u}{\partial y} = \frac{\partial u}{\partial r}\frac{\partial r}{\partial y} + \frac{\partial u}{\partial \varphi}\frac{\partial \varphi}{\partial y},$$

or

$$\frac{\partial u}{\partial x}=\frac{\partial u}{\partial r}\frac{x}{\sqrt{x^2+y^2}}-\frac{\partial u}{\partial \varphi}\frac{y}{x^2+y^2}=\cos\varphi\frac{\partial u}{\partial r}-\frac{1}{r}\sin\varphi\frac{\partial u}{\partial \varphi},$$

$$\frac{\partial u}{\partial y}=\frac{\partial u}{\partial r}\frac{y}{\sqrt{x^2+y^2}}+\frac{\partial u}{\partial \varphi}\frac{x}{x^2+y^2}=\sin\varphi\frac{\partial u}{\partial r}+\frac{1}{r}\cos\varphi\frac{\partial u}{\partial \varphi}.$$

Hence, we have the following formulae:

$$\frac{\partial}{\partial x}=\cos\varphi\frac{\partial}{\partial r}-\frac{1}{r}\sin\varphi\frac{\partial}{\partial \varphi}$$

$$\frac{\partial}{\partial y}=\sin\varphi\frac{\partial}{\partial r}+\frac{1}{r}\cos\varphi\frac{\partial}{\partial \varphi}.$$

To find the Cartesian slope of the tangent line to a polar curve $r(\varphi)$ at any given point, the curve is first expressed as a system of parametric equations.

$$x=r(\varphi)\cos\varphi$$

$$y=r(\varphi)\sin\varphi$$

Differentiating both equations with respect to φ yields

$$\frac{dx}{d\varphi}=r'(\varphi)\cos\varphi-r(\varphi)\sin\varphi$$

$$\frac{dy}{d\varphi}=r'(\varphi)\sin\varphi+r(\varphi)\cos\varphi.$$

Dividing the second equation by the first yields the Cartesian slope of the tangent line to the curve at the point $(r(\varphi), \varphi)$:

$$\frac{dy}{dx}=\frac{r'(\varphi)\sin\varphi+r(\varphi)\cos\varphi}{r'(\varphi)\cos\varphi-r(\varphi)\sin\varphi}.$$

Integral Calculus (Arc Length)

The arc length (length of a line segment) defined by a polar function is found by the integration over the curve $r(\varphi)$. Let L denote this length along the curve starting from points A through to point B, where these points correspond to $\varphi = a$ and $\varphi = b$ such that $0 < b - a < 2\pi$. The length of L is given by the following integral

$$L=\int_a^b\sqrt{[r(\varphi)]^2+\left[\tfrac{dr(\varphi)}{d\varphi}\right]^2}\,d\varphi$$

Integral Calculus (Area)

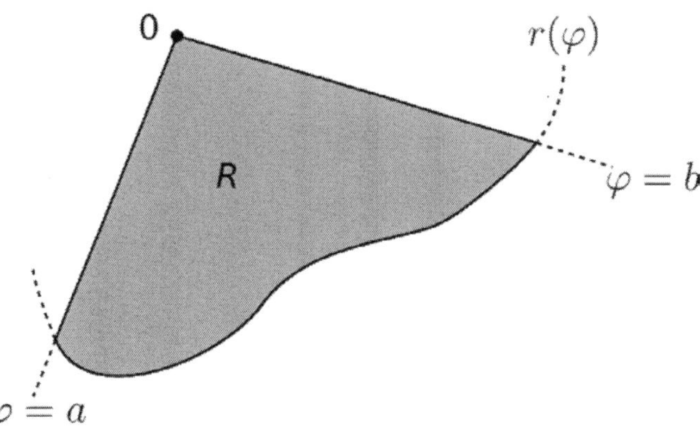

Figure: *The integration region R is bounded by the curve r(φ) and the rays φ = a and φ = b.*

Let R denote the region enclosed by a curve $r(\varphi)$ and the rays $\varphi = a$ and $\varphi = b$, where $0 < b - a \leq 2\pi$. Then, the area of R is

$$\frac{1}{2}\int_a^b [r(\varphi)]^2 \, d\varphi.$$

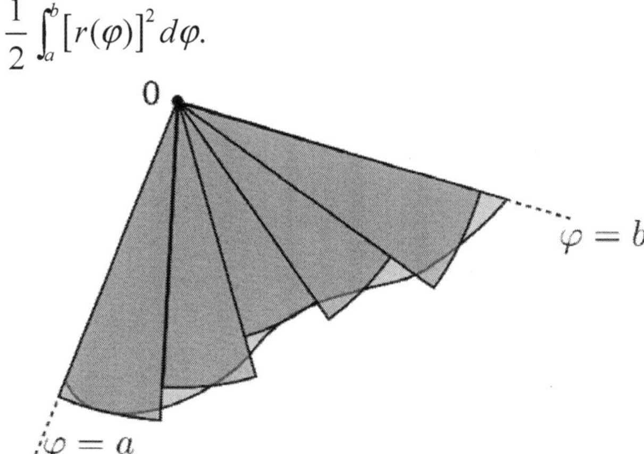

Figure: *The region R is approximated by n sectors (here, n = 5).*

This result can be found as follows. First, the interval $[a, b]$ is divided into n subintervals, where n is an arbitrary positive integer. Thus $\Delta\varphi$, the length of each subinterval, is equal to b " a (the total length of the interval), divided by n, the number of subintervals. For each subinterval $i = 1, 2, \ldots, n$, let φ_i be the midpoint of the subinterval, and construct a sector with the center at the pole, radius $r(\varphi_i)$, central angle $\Delta\varphi$ and arc length $r(\varphi_i)\Delta\varphi$. The area of each constructed sector is therefore equal to

$$[r(\varphi_i)]^2 \pi \cdot \frac{\Delta\varphi}{2\pi} = \frac{1}{2}[r(\varphi_i)]^2 \Delta\varphi.$$

Hence, the total area of all of the sectors is

$$\sum_{i=1}^{n} \tfrac{1}{2} r(\varphi_i)^2 \Delta\varphi.$$

As the number of subintervals n is increased, the approximation of the area continues to improve. In the limit as $n \to \infty$, the sum becomes the Riemann sum for the above integral.

A mechanical device that computes area integrals is the planimeter, which measures the area of plane figures by tracing them out: this replicates integration in polar coordinates by adding a joint so that the 2-element linkage effects Green's theorem, converting the quadratic polar integral to a linear integral.

Generalization

Using Cartesian coordinates, an infinitesimal area element can be calculated as $dA = dx\,dy$. The substitution rule for multiple integrals states that, when using other coordinates, the Jacobian determinant of the coordinate conversion formula has to be considered:

$$J = \det\frac{\partial(x,y)}{\partial(r,\varphi)} = \begin{vmatrix} \frac{\partial x}{\partial r} & \frac{\partial x}{\partial \varphi} \\ [8pt]\frac{\partial y}{\partial r} & \frac{\partial y}{\partial \varphi} \end{vmatrix} = \begin{vmatrix} \cos\varphi & -r\sin\varphi \\ \sin\varphi & r\cos\varphi \end{vmatrix} = r\cos^2\varphi + r\sin^2\varphi = r.$$

Hence, an area element in polar coordinates can be written as

$$dA = dx dy = J\,dr\,d\varphi = r\,dr\,d\varphi.$$

Now, a function that is given in polar coordinates can be integrated as follows:

$$\iint_R f(x,y)dA = \int_a^b \int_0^{r(\varphi)} f(r,\varphi) r\,dr\,d\varphi.$$

Here, R is the same region as above, namely, the region enclosed by a curve $r(\varphi)$ and the rays $\varphi = a$ and $\varphi = b$.

The formula for the area of R mentioned above is retrieved by taking f identically equal to 1. A more surprising application of this result yields the Gaussian integral

$$\int_{-\infty}^{\infty} e^{-x^2}\,dx = \sqrt{\pi}.$$

Vector Calculus

Vector calculus can also be applied to polar coordinates. For a planar motion, let r be the position vector $(r\cos(\varphi), r\sin(\varphi))$, with r and φ depending on time t.

We define the unit vectors

$$\hat{\mathrm{r}} = (\cos(\varphi), \sin(\varphi))$$

in the direction of r and

$$\hat{\varphi} = (-\sin(\varphi), \cos(\varphi)) = \hat{\mathrm{k}} \times \hat{\mathrm{r}},$$

in the plane of the motion perpendicular to the radial direction, where $\hat{\mathbf{k}}$ is a unit vector normal to the plane of the motion.

Then

$$\mathrm{r} = (x, y) = r(\cos\varphi, \sin\varphi) = r\hat{\mathrm{r}},$$

$$\dot{\mathrm{r}} = (\dot{x}, \dot{y}) = \dot{r}(\cos\varphi, \sin\varphi) + r\dot{\varphi}(-\sin\varphi, \cos\varphi) = \dot{r}\hat{\mathrm{r}} + r\dot{\varphi}\hat{\varphi},$$

$$\ddot{\mathbf{r}} = (\ddot{x}, \ddot{y}) = \ddot{r}(\cos\varphi, \sin\varphi) + 2\dot{r}\dot{\varphi}(-\sin\varphi, \cos\varphi) + r\ddot{\varphi}(-\sin\varphi, \cos\varphi) - r\dot{\varphi}^2(\cos\varphi, \sin\varphi) =$$

$$\left(\ddot{r} - r\dot{\varphi}^2\right)\hat{\mathrm{r}} + \left(r\ddot{\varphi} + 2\dot{r}\dot{\varphi}\right)\hat{\varphi} = (\ddot{r} - r\dot{\varphi}^2)\hat{\mathrm{r}} + \frac{1}{r}\frac{d}{dt}\left(r^2\dot{\varphi}\right)\hat{\varphi}$$

Centrifugal and Coriolis Terms

The term $r\dot{\varphi}^2$ is sometimes referred to as the *centrifugal term*, and the term $2\dot{r}\dot{\varphi}$ as the *Coriolis term*. Although these equations bear some resemblance in form to the centrifugal and Coriolis effects found in rotating reference frames, nonetheless these are not the same things. For example, the physical centrifugal and Coriolis forces appear only in non-inertial frames of reference. In contrast, these terms that appear when acceleration is expressed in polar coordinates are a mathematical consequence of differentiation; these terms appear wherever polar coordinates are used. In particular, these terms appear even when polar coordinates are used in inertial frames of reference, where the physical centrifugal and Coriolis forces never appear.

Co-Rotating Frame

For a particle in planar motion, one approach to attaching physical significance to these terms is based on the concept of an instantaneous *co-rotating frame of reference.* To define a co-rotating frame, first an origin is selected from which the distance $r(t)$ to the particle is defined. An axis of rotation is set up that is perpendicular to the plane of motion of the particle, and passing through this origin. Then, at the selected moment t, the rate of rotation of the co-rotating frame Ω is made to match the rate of rotation of the particle about this axis, $d\varphi/dt$. Next, the terms in the acceleration in the inertial frame are related to those in the co-rotating frame. Let the location of the particle in the inertial frame be $(r(t), \varphi(t))$, *and in the co-rotating frame be* (r(t), φ2 (t)). *Because the co-rotating frame rotates at the same rate as the particle,* dφ2 /dt = 0. *The fictitious centrifugal force in the co-rotating frame is* $mr\Omega^2$, radially outward. The velocity of the particle in the co-rotating frame also is radially outward, because *d*φ2 /*dt* = 0. The *fictitious Coriolis force* therefore has a value "2*m*(*dr*/*dt*)Ω, pointed in the direction of increasing φ only. Thus, using these forces in Newton's second law we find:

$$F + F_{cf} + F_{Cor} = m\ddot{r},$$

where over dots represent time differentiations, and F is the net real force (as opposed to the fictitious forces). In terms of components, this vector equation becomes:

$$F_r + mr\Omega^2 = m\ddot{r}$$

$$F_\varphi - 2m\dot{r}\Omega = mr\ddot{\varphi},$$

which can be compared to the equations for the inertial frame:

$$F_r = m\ddot{r} - mr\dot{\varphi}^2$$

$$F_\varphi = mr\ddot{\varphi} + 2m\dot{r}\dot{\varphi}.$$

This comparison, plus the recognition that by the definition of the co-rotating frame at time *t* it has a rate of rotation Ω = *d*φ/*dt*, shows that we can interpret the terms in the acceleration (multiplied by the mass of the particle) as found in the inertial frame as the negative of the centrifugal and Coriolis forces that would be seen in the instantaneous, non-inertial co-rotating frame.

For general motion of a particle (as opposed to simple circular motion), the centrifugal and Coriolis forces in a particle's frame of reference commonly are referred to the instantaneous osculating circle of its motion, not to a fixed center of polar coordinates.

Connection to Spherical and Cylindrical Coordinates

The polar coordinate system is extended into three dimensions with two different coordinate systems, the cylindrical and spherical coordinate system.

Applications

Polar coordinates are two-dimensional and thus they can be used only where point positions lie on a single two-dimensional plane. They are most appropriate in any context where the phenomenon being considered is inherently tied to direction and length from a center point.

For instance, the examples above show how elementary polar equations suffice to define curves—such as the Archimedean spiral—whose equation in the Cartesian coordinate system would be much more intricate. Moreover, many physical systems—such as those concerned with bodies moving around a central point or with phenomena originating from a central point—are simpler and more intuitive to model using polar coordinates. The initial motivation for the introduction of the polar system was the study of circular and orbital motion.

Position and Navigation

Polar coordinates are used often in navigation, as the destination or direction of travel can be given as an angle and distance from the object being considered. For instance, aircraft use a slightly modified version of the polar coordinates for navigation. In this system, the one generally used for any sort of navigation, the 0° ray is generally called heading 360, and the angles continue in a clockwise direction, rather than counterclockwise, as in the mathematical system. Heading 360 corresponds to magnetic north, while headings 90, 180, and 270 correspond to magnetic east, south, and west, respectively. Thus, an aircraft travelling 5 nautical miles due east will be travelling 5 units at heading 90 (read zero-niner-zero by air traffic control).

Modelling

Systems displaying radial symmetry provide natural settings for the polar coordinate system, with the central point acting as the pole. A prime example of this usage is the groundwater flow equation when applied to radially symmetric wells. Systems with a radial force are also good candidates for the use of the polar coordinate system. These systems include gravitational fields, which obey the inverse-square law, as well as systems with point sources, such as radio antennas.

Radially asymmetric systems may also be modelled with polar coordinates. For example, a microphone's pickup pattern illustrates its proportional response to an incoming sound from a given direction, and these patterns can be represented as polar curves. The curve for a standard cardioid microphone, the most common unidirectional microphone, can be represented as $r = 0.5 + 0.5\sin(\varphi)$ at its target design frequency. The pattern shifts toward omnidirectionality at lower frequencies.

Chapter 3

Spherical Coordinate Systems

In mathematics, a spherical coordinate system is a coordinate system for three-dimensional space where the position of a point is specified by three numbers: the radial distance of that point from a fixed origin, its polar angle measured from a fixed zenith direction, and the azimuth angle of its orthogonal projection on a reference plane that passes through the origin and is orthogonal to the zenith, measured from a fixed reference direction on that plane.

The radial distance is also called the radius or radial coordinate. The polar angle may be called co-latitude, zenith angle, normal angle, or inclination angle.

The use of symbols and the order of the coordinates differs between sources. In one system frequently encountered in physics (r, θ, φ) gives the radial distance, polar angle, and azimuthal angle, whereas in another system used in many mathematics books (r, θ, φ) gives the radial distance, azimuthal angle, and polar angle. In both systems ρ is often used instead of r. Other conventions are also used, so great care needs to be taken to check which one is being used.

A number of different spherical coordinate systems following other conventions are used outside mathematics. In a geographical coordinate system positions are measured in latitude, longitude and height or altitude. There are a number of different celestial coordinate systems based on different fundamental planes and with different terms for the various coordinates. The spherical coordinate systems used in mathematics normally use radians rather than degrees and measure the azimuthal angle counter-clockwise rather than clockwise[further explanation needed]. The inclination angle is often replaced by the elevation angle measured from the reference plane. Elevation angle of zero is at the horizon.

The concept of spherical coordinates can be extended to higher-dimensional spaces and are then referred to as hyperspherical coordinates.

Conventions

Several different conventions exist for representing the three coordinates, and for the order in which they should be written. The use of (r, θ, φ) to denote radial distance, inclination (or elevation), and azimuth, respectively, is common practice in physics, and is specified by ISO standard 80000-2 :2009.

However, some authors (including mathematicians) use φ for inclination (or elevation) and θ for azimuth, which "provides a logical extension of the usual polar coordinates notation". Some authors may also list the azimuth before the inclination (or elevation), and/or use *ñ* (rho) instead of r for radial distance. Some combinations of these choices result in a left-handed coordinate system. The standard convention (r, θ, φ) conflicts with the usual notation for the two-dimensional polar coordinates, where θ is often used for the azimuth. It may also conflict with the notation used for three-dimensional cylindrical coordinates.

The angles are typically measured in degrees (°) or radians (rad), where 360° = 2π rad. Degrees are most common in geography, astronomy, and engineering, whereas radians are commonly used in mathematics and theoretical physics. The unit for radial distance is usually determined by the context.

When the system is used for physical three-space, it is customary to use positive sign for azimuth angles that are measured in the counter-clockwise sense from the reference direction on the reference plane, as seen from the zenith side of the plane. This convention is used, in particular, for geographical coordinates, where the "zenith" direction is north and positive azimuth (longitude) angles are measured eastwards from some prime meridian.

Major Conventions

Coordinates	*Corresponding Local Geographical Directions*	*Right/Left-Handed*
		right
		right
		left

Note: easting (E), northing (N), upwardness (U). Local azimuth angle would be measured, e.g., counterclockwise from S to in the case of .

Unique Coordinates

Any spherical coordinate triplet (r, θ, φ) specifies a single point of three-dimensional space. On the other hand, every point has infinitely many equivalent spherical coordinates. One can add or subtract any number of full turns to either angular measure without changing the angles themselves, and therefore without changing the point. It is also convenient, in many contexts, to allow negative radial distances, with the convention that $(-r, \theta, \varphi)$ is equivalent to $(r, \theta + 180°, \varphi)$ for any r, θ, and φ. Moreover, $(r, -\theta, \varphi)$ is equivalent to $(r, \theta, \varphi + 180°)$.

If it is necessary to define a unique set of spherical coordinates for each point, one may restrict their ranges. A common choice is:

$$r \geq 0$$

$$0° \leq \theta \leq 180° \ (\pi \text{ rad})$$

$$0° \leq \varphi < 360° \ (2\pi \text{ rad})$$

However, the azimuth φ is often restricted to the interval $(-180°, +180°]$, or $(-\pi, +\pi]$ in radians, instead of $[0, 360°)$. This is the standard convention for geographic longitude.

The range $[0°, 180°]$ for inclination is equivalent to $[-90°, +90°]$ for elevation (latitude).

Even with these restrictions, if θ is zero or 180° (elevation is 90° or −90°) then the azimuth angle is arbitrary; and if r is zero, both azimuth and inclination/elevation are arbitrary. To make the coordinates unique, one can use the convention that in these cases the arbitrary coordinates are zero.

Plotting

To plot a point from its spherical coordinates (r, θ, φ), where θ is inclination, move r units from the origin in the zenith direction, rotate by θ about the origin towards the azimuth reference direction, and rotate by φ about the zenith in the proper direction.

Applications

The geographic coordinate system uses the azimuth and elevation of the spherical coordinate system to express locations on Earth, calling them respectively longitude and latitude. Just as the two-dimensional Cartesian coordinate system is useful on the plane, a two-dimensional spherical coordinate system is useful on the surface of a sphere. In this system, the sphere is taken as a unit sphere, so the radius is unity and can generally be ignored. This simplification can also be very useful when dealing with objects such as rotational matrices.

Spherical coordinates are useful in analyzing systems that have some degree of symmetry about a point, such as volume integrals inside a sphere, the potential energy field surrounding a concentrated mass or charge, or global weather simulation in a planet's atmosphere. A sphere that has the Cartesian equation $x^2 + y^2 + z^2 = c^2$ has the simple equation $r = c$ in spherical coordinates.

Two important partial differential equations that arise in many physical problems, Laplace's equation and the Helmholtz equation, allow a separation of variables in spherical coordinates. The angular portions of the solutions to such equations take the form of spherical harmonics.

Another application is ergonomic design, where r is the arm length of a stationary person and the angles describe the direction of the arm as it reaches out.

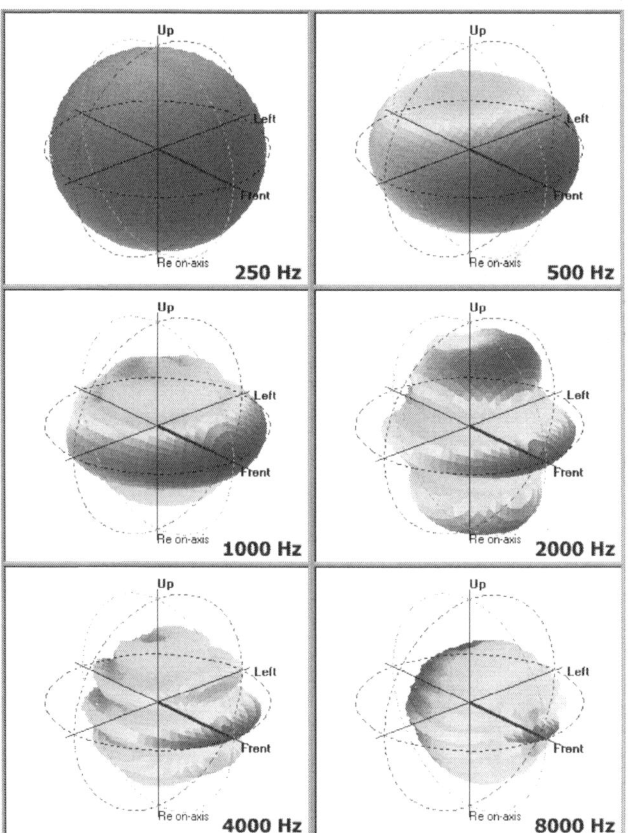

Figure: *The output pattern of an industrial loudspeaker shown using spherical polar plots taken at six frequencies*

Three dimensional modelling of loudspeaker output patterns can be used to predict their performance. A number of polar plots are required, taken at a wide selection of frequencies, as the pattern changes greatly with frequency. Polar plots help to show that many loudspeakers tend toward omnidirectionality at lower frequencies.

The spherical coordinate system is also commonly used in 3D game development to rotate the camera around the player's position.

Coordinate System Conversions

As the spherical coordinate system is only one of many three-dimensional coordinate systems, there exist equations for converting coordinates between the spherical coordinate system and others.

Cartesian Coordinates

The spherical coordinates of a point in the ISO convention (*radius* r, *inclination* θ, *azimuth* φ) can be obtained from its Cartesian coordinates (x, y, z) by the formulae

$$r = \sqrt{x^2 + y^2 + z^2}$$

$$\theta = \arccos\left(\frac{z}{\sqrt{x^2 + y^2 + z^2}}\right)$$

$$\varphi = \arctan\left(\frac{y}{x}\right)$$

The inverse tangent denoted in $\varphi = \arctan(y/x)$ must be suitably defined, taking into account the correct quadrant of (x,y).

Alternatively, the conversion can be considered as two sequential rectangular to polar conversions: the first in the Cartesian x–y plane from (x,y) to (R,φ), where R is the projection of r onto the x–y plane, and the second in the Cartesian z–R plane from (z,R) to (r,θ). The correct quadrants for φ and θ are implied by the correctness of the planar rectangular to polar conversions.

These formulae assume that the two systems have the same origin, that the spherical reference plane is the Cartesian x–y plane, that θ is inclination from the z direction, and that the azimuth angles are measured from the Cartesian x axis (so that the y axis has $\varphi = +90°$). If θ measures elevation from the reference plane instead of inclination from the zenith the across above becomes an arcsin, and the $\cos\theta$ and $\sin\theta$ below become switched.

Conversely, the Cartesian coordinates may be retrieved from the spherical coordinates (*radius r*, *inclination* θ, *azimuth* φ), where $r \in [0, \infty)$, $\theta \in [0, \pi]$, $\varphi \in [0, 2\pi)$, by:

$$x = r\sin\theta\cos\varphi$$

$$y = r\sin\theta\sin\varphi$$

$$z = r\cos\theta$$

Geographic Coordinates

To a first approximation, the geographic coordinate system uses elevation angle (latitude) in degrees north of the equator plane, in the range $-90° \leq \varphi \leq 90°$, instead of inclination. Latitude is either geocentric latitude, measured at the Earth's center and designated variously by ø, *q*, φ2 , φ_c, φ_g or geodetic latitude, measured by the observer's local vertical, and commonly designated φ. The azimuth angle (longitude), commonly denoted by λ, is measured in degrees east or west from some conventional reference meridian (most commonly the IERS Reference Meridian), so its domain is $-180° \leq \lambda \leq 180°$. For positions on the Earth or other solid celestial body, the reference plane is usually taken to be the plane perpendicular to the axis of rotation.

The polar angle, which is 90° minus the latitude and ranges from 0 to 180°, is called colatitude in geography.

Instead of the radial distance, geographers commonly use altitude above some reference surface, which may be the sea level or "mean" surface level for planets without liquid oceans. The radial distance *r* can be computed from the altitude by adding the mean radius of the planet's reference surface, which is approximately 6,360 ± 11 km for Earth.

However, modern geographical coordinate systems are quite complex, and the positions implied by these simple formulae may be wrong by several kilometers. The precise standard meanings of latitude, longitude and altitude are currently defined by the World Geodetic System (WGS), and take into account the flattening of the Earth at the poles (about 21 km) and many other details.

In astronomy there are a series of spherical coordinate systems that measure the elevation angle from different fundamental planes. These reference planes are the observer's horizon, the celestial equator (defined by the Earth's rotation), the plane of the ecliptic (defined by Earth's orbit around the sun), and the galactic equator (defined by the rotation of the galaxy).

Cylindrical Coordinates

Cylindrical coordinates (*radius ñ, azimuth* φ, *elevation z*) may be converted into spherical coordinates (*radius r, inclination* θ, *azimuth* φ), by the formulas

$$r = \sqrt{\rho^2 + z^2}$$

$$\theta = \arctan(\rho / z) = \arccos\left(\frac{z}{\sqrt{\rho^2 + z^2}}\right)$$

$$\varphi = \varphi$$

Conversely, the spherical coordinates may be converted into cylindrical coordinates by the formulae

$$\rho = r\sin\theta$$

$$\varphi = \varphi$$

$$z = r\cos\theta$$

These formulae assume that the two systems have the same origin and same reference plane, measure the azimuth angle φ in the same sense from the same axis, and that the spherical angle θ is inclination from the cylindrical *z* axis.

Kinematics

In spherical coordinates the position of a point is written,

$$\mathbf{r} = r\hat{\mathbf{r}}$$

its velocity is then,

$$\mathbf{v} = \dot{r}\hat{\mathbf{r}} + r\dot{\theta}\hat{\boldsymbol{\theta}} + r\dot{\varphi}\sin\theta\hat{\boldsymbol{\varphi}}$$

and its acceleration is,

$$\mathbf{a} = \left(\ddot{r} - r\dot{\theta}^2 - r\dot{\varphi}^2\sin^2\theta\right)\hat{\mathbf{r}}$$

$$+\left(r\ddot{\theta} + 2\dot{r}\dot{\theta} - r\dot{\varphi}^2\sin\theta\cos\theta\right)\hat{\boldsymbol{\theta}}$$

$$+\left(r\ddot{\varphi}\sin\theta + 2\dot{r}\dot{\varphi}\sin\theta + 2r\dot{\theta}\dot{\varphi}\cos\theta\right)\hat{\boldsymbol{\varphi}}$$

In the case of a constant φ or $\theta = \frac{\pi}{2}$, this reduces to vector calculus in polar coordinates.

Divergence Theorem and Stokes Theorem

In vector calculus, the divergence theorem, also known as Gauss's theorem or Ostrogradsky's theorem, is a result that relates the flow (that is, flux) of a vector field through a surface to the behaviour of the vector field inside the surface.

More precisely, the divergence theorem states that the outward flux of a vector field through a closed surface is equal to the volume integral of the divergence over the region inside the surface. Intuitively, it states that *the sum of all sources minus the sum of all sinks gives the net flow out of a region.*

The divergence theorem is an important result for the mathematics of engineering, in particular in electrostatics and fluid dynamics.

In physics and engineering, the divergence theorem is usually applied in three dimensions. However, it generalizes to any number of dimensions. In one dimension, it is equivalent to the fundamental theorem of calculus. In two dimensions, it is equivalent to Green's theorem.

The theorem is a special case of the more general Stokes' theorem.

Intuition

If a fluid is flowing in some area, then the rate at which fluid flows out of a certain region within that area can be calculated by adding up the sources inside the region and subtracting the sinks. The fluid flow is represented by a vector field, and the vector field's divergence at a given point describes the strength of the source or sink there. So, integrating the field's divergence over the interior of the region should equal the integral of the vector field over the region's boundary. The divergence theorem says that this is true.

The divergence theorem is thus a conservation law which states that the volume total of all sinks and sources, that is the volume integral of the divergence, is equal to the net flow across the volume's boundary.

Mathematical Statement

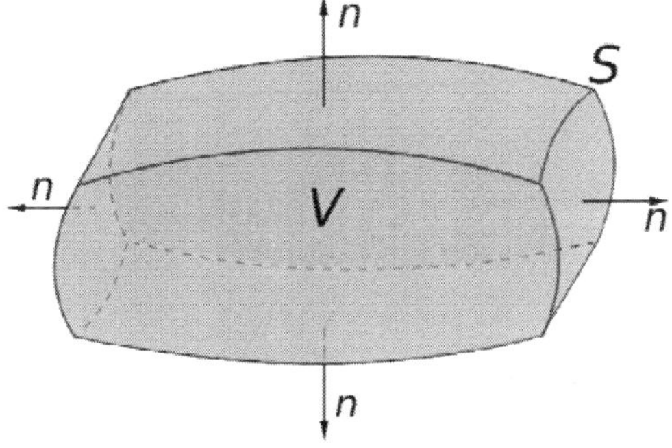

Figure: *A region* V *bounded by the surface* S = ∂V *with the surface normal n*

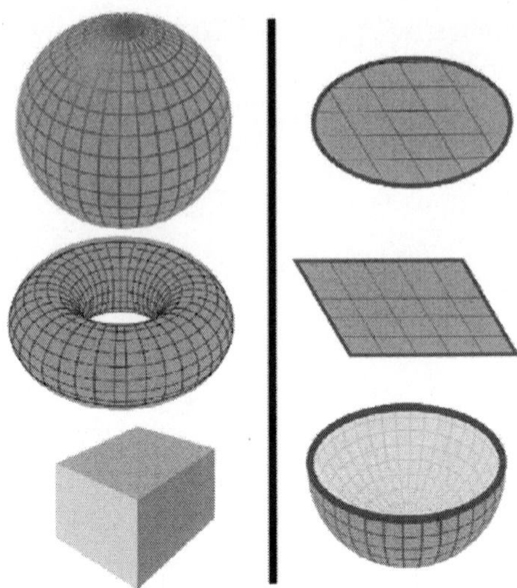

Figure: *The divergence theorem can be used to calculate a flux through a closed surface that fully encloses a volume, like any of the surfaces on the left. It can not directly be used to calculate the flux through surfaces with boundaries, like those on the right. (Surfaces are blue, boundaries are red.)*

Suppose V is a subset of $\mathbb{R}^n$ (in the case of $n = 3$, V represents a volume in 3D space) which is compact and has a piecewise smooth boundary S (also indicated with $\partial V = S$). If F is a continuously differentiable vector field defined on a neighbourhood of V, then we have:

$$\iiint_V (\nabla\cdot\mathbf{F})dV = \oiint \mathbf{s}(\mathbf{F}\cdot\mathbf{n})dS.$$

The left side is a volume integral over the volume V, the right side is the surface integral over the boundary of the volume V. The closed manifold ∂V is quite generally the boundary of V oriented by outward-pointing normals, and n is the outward pointing unit normal field of the boundary ∂V. (dS may be used as a shorthand for ndS.) The symbol within the two integrals stresses once more that ∂V is a *closed* surface. In terms of the intuitive description above, the left-hand side of the equation represents the total of the sources in the volume V, and the right-hand side represents the total flow across the boundary ∂V.

Corollaries

By applying the divergence theorem in various contexts, other useful identities can be derived (cf. vector identities).

- Applying the divergence theorem to the product of a scalar function g and a vector field F, the result is

$$\iiint_V \left[\mathbf{F}\cdot(\nabla g) + g(\nabla\cdot\mathbf{F})\right]dV = \oiint_{sg} F \cdot dS$$

A special case of this is F = ∇f, in which case the theorem is the basis for Green's identities.

- Applying the divergence theorem to the cross-product of two vector fields F × G, the result is

$$\iiint_V \left[\mathbf{G}\cdot(\nabla\times\mathbf{F}) - \mathbf{F}\cdot(\nabla\times\mathbf{G})\right] dV = \oiint_S (F\times G)\cdot dS.$$

- Applying the divergence theorem to the product of a scalar function, f, and a non-zero constant vector c, the following theorem can be proven:

$$\iiint_V \mathbf{c}\cdot\nabla f\, dV = \oiint_S (\mathrm{c}f)\cdot d\mathrm{S} - \iiint_V f(\nabla\cdot\mathrm{c})dV.$$

- Applying the divergence theorem to the cross-product of a vector field F and a non-zero constant vector c, the following theorem can be proven:

$$\iiint_V \mathrm{c}\cdot(\nabla\times\mathrm{F})dV = \oiint_S (\mathrm{c}f)\cdot d\mathrm{S} - \iiint_V f(\nabla\cdot\mathrm{c})dV.$$

Example

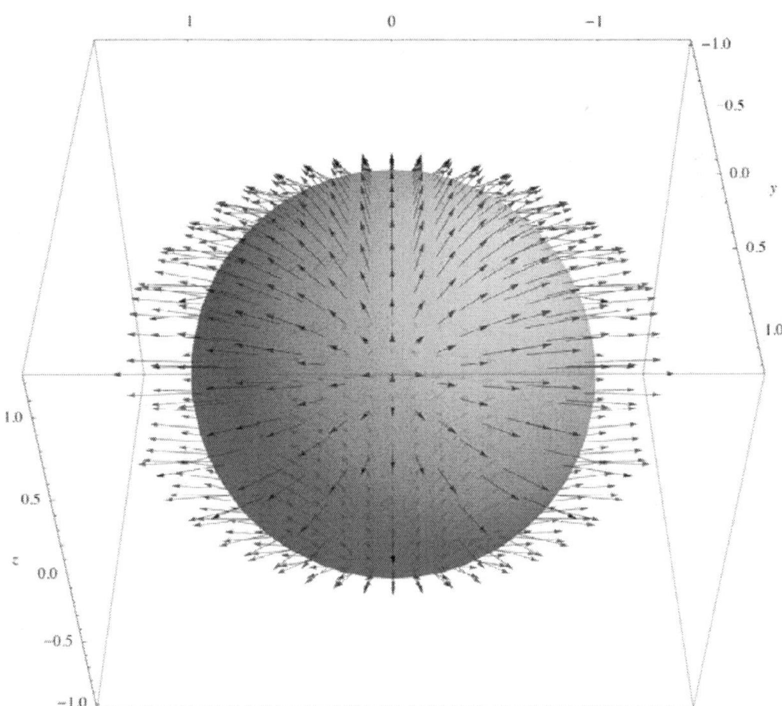

Figure: *The vector field corresponding to the example shown. Note, vectors may point into or out of the sphere.*

Suppose we wish to evaluate

$$\oiint S\mathrm{F}\cdot\mathrm{n}dS,$$

where S is the unit sphere defined by

$$S=\left\{x,y,z\in\mathbb{R}^3 : x^2+y^2+z^2=1\right\}.$$

and F is the vector field

$$\mathrm{F}=2x\mathrm{i}+y^2\mathrm{j}+z^2\mathrm{k}.$$

The direct computation of this integral is quite difficult, but we can simplify the derivation of the result using the divergence theorem, because the divergence theorem says that the integral is equal to:

$$\iiint_W(\nabla\cdot\mathbf{F})dV=2\iiint_W(1+y+z)dV=2\iiint_W dV+2\iiint_W ydV+2\iiint_W zdV.$$

where W is the unit ball:

$$W=\left\{x,y,z\in\mathbb{R}^3 : x^2+y^2+z^2\leq 1\right\}.$$

Since the function y is positive in one hemisphere of W and negative in the other, in an equal and opposite way, its total integral over W is zero. The same is true for z:

$$\iiint_W ydV=\iiint_W zdV=0.$$

Therefore,

$$\oiint S$$

because the unit ball W has volume 4ð/3.

Applications

Differential Form and Integral Form of Physical Laws: As a result of the divergence theorem, a host of physical laws can be written in both a differential form (where one quantity is the divergence of another) and an integral form (where the flux of one quantity through a closed surface is equal to another quantity). Three examples are Gauss's law (in electrostatics), Gauss's law for magnetism, and Gauss's law for gravity.

Continuity Equations

Continuity equations offer more examples of laws with both differential and integral forms, related to each other by the divergence theorem. In fluid dynamics, electromagnetism, quantum mechanics, relativity theory, and a number of other fields, there are continuity equations that describe the conservation of mass, momentum, energy, probability, or other quantities. Generically, these equations state that the divergence of the flow of the conserved quantity is equal to the distribution of *sources* or *sinks* of that quantity. The divergence theorem states that any such continuity equation can be written in a differential form (in terms of a divergence) and an integral form (in terms of a flux).

Inverse-Square Laws

Any inverse-square law can instead be written in a *Gauss' law*-type form (with a differential and integral form, as described above). Two examples are Gauss' law (in electrostatics), which follows from the inverse-square Coulomb's law, and Gauss' law for gravity, which follows from the inverse-square Newton's law of universal gravitation. The derivation of the Gauss' law-type equation from the inverse-square formulation (or vice-versa) is exactly the same in both cases.

History

The theorem was first discovered by Lagrange in 1762, then later independently rediscovered by Gauss in 1813, by Ostrogradsky, who also gave the first proof of the general theorem, in 1826, by Green in 1828, etc. Subsequently, variations on the divergence theorem are correctly called Ostrogradsky's theorem, but also commonly Gauss's theorem, or Green's theorem.

Generalizations

Multiple Dimensions: One can use the general Stokes' Theorem to equate the n-dimensional volume integral of the divergence of a vector field F over a region U to the $(n-1)$-dimensional surface integral of F over the boundary of U:

$$\int_U \nabla\cdot \mathrm{F}\, dV_n = \oint_{\partial U} \mathrm{F}\cdot \mathrm{n}\, dS_{n-1}$$

This equation is also known as the Divergence theorem.

When $n = 2$, this is equivalent to Green's theorem.

When $n = 1$, it reduces to the Fundamental theorem of calculus.

Tensor Fields

Writing the theorem in Einstein notation:

$$\iiint_V \frac{\partial \mathbf{F}_i}{\partial x_i} dV = \oiint\; S\mathbf{F}_i n_i\, dS$$

suggestively, replacing the vector field F with a rank-n tensor field T, this can be generalized to:

$$\iiint_V \frac{\partial T_{i_1 i_2 \cdots i_q \cdots i_n}}{\partial x_{i_q}} dV = \oiint_S T_{i_1 i_2 \cdots i_q \cdots i_n} n_{i_q} \, dS.$$

where on each side, tensor contraction occurs for at least one index. This form of the theorem is still in 3d, each index takes values 1, 2, and 3. It can be generalized further still to higher (or lower) dimensions.

Stokes' Theorem

In differential geometry, Stokes' theorem (also called the generalized Stokes' theorem) is a statement about the integration of differential forms on manifolds, which both simplifies and generalizes several theorems from vector calculus. Stokes' theorem says that the integral of a differential form ω over the boundary of some orientable manifold Ω is equal to the integral of its exterior derivative dω over the whole of Ω, i.e.

$$\int_{\partial\Omega} \omega = \int_{\Omega} \mathrm{d}\omega.$$

This modern form of Stokes' theorem is a vast generalization of a classical result first discovered by Lord Kelvin, who communicated it to George Stokes in a letter dated July 2, 1850. Stokes set the theorem as a question on the 1854 Smith's Prize exam, which led to the result bearing his name. This classical Kelvin–Stokes theorem relates the surface integral of the curl of a vector field F over a surface Σ in Euclidean three-space to the line integral of the vector field over its boundary ∂Σ:

$$\iint_{\Sigma} \nabla \times \mathbf{F} \cdot \mathrm{d}\mathbf{\Sigma} = \oint_{\partial\Sigma} \mathbf{F} \cdot \mathrm{d}\mathbf{r}.$$

This classical statement, along with the classical divergence theorem, fundamental theorem of calculus, and Green's theorem are simply special cases of the general formulation stated above.

The fundamental theorem of calculus states that the integral of a function f over the interval $[a, b]$ can be calculated by finding an antiderivative F of f :

$$\int_a^b f(x)\mathrm{d}x = F(b) - F(a).$$

Stokes' theorem is a vast generalization of this theorem in the following sense.

- By the choice of F, $\frac{\mathrm{d}F}{\mathrm{d}x} = f(x)$. In the parlance of differential forms, this is saying that $f(x)$ dx is the exterior derivative of the

0-form, i.e. function, F: in other words, that d$F = f$ dx. The general Stokes theorem applies to higher differential forms ω instead of just 0-forms such as F.

- A closed interval $[a, b]$ is a simple example of a one-dimensional manifold with boundary. Its boundary is the set consisting of the two points a and b. Integrating f over the interval may be generalized to integrating forms on a higher-dimensional manifold. Two technical conditions are needed: the manifold has to be orientable, and the form has to be compactly supported in order to give a well-defined integral.
- The two points a and b form the boundary of the open interval. More generally, Stokes' theorem applies to oriented manifolds M with boundary. The boundary ∂M of M is itself a manifold and inherits a natural orientation from that of the manifold. For example, the natural orientation of the interval gives an orientation of the two boundary points. Intuitively, a inherits the opposite orientation as b, as they are at opposite ends of the interval. So, "integrating" F over two boundary points a, b is taking the difference $F(b)\ \partial\ F(a)$.

In even simpler terms, one can consider that points can be thought of as the boundaries of curves, that is as 0-dimensional boundaries of 1-dimensional manifolds. So, just as one can find the value of an integral (f dx = dF) over a 1-dimensional manifolds ($[a, b]$) by considering the anti-derivative (F) at the 0-dimensional boundaries ($[a, b]$), one can generalize the fundamental theorem of calculus, with a few additional caveats, to deal with the value of integrals (dω) over n-dimensional manifolds (Ω) by considering the anti-derivative (ω) at the $(n - 1)$-dimensional boundaries (dΩ) of the manifold.

So the fundamental theorem reads:

$$\int_{[a,b]} f(x)\mathrm{d}x = \int_{[a,b]} \mathrm{d}F = \int_{\{a\}^- \cup \{b\}^+} F = F(b) - F(a).$$

General Formulation

Let Ω be an oriented smooth manifold of dimension n and let α be an n-differential form that is compactly supported on Ω. First, suppose that α is compactly supported in the domain of a single, oriented coordinate chart $\{U, \varphi\}$. In this case, we define the integral of α over Ω as

$$\int_\Omega \alpha = \int_{\phi(U)} \left(\phi^{-1}\right)^* \alpha,$$

i.e., via the pullback of α to R^n.

More generally, the integral of α over Ω is defined as follows: Let $\{\varnothing_i\}$ be a partition of unity associated with a locally finite cover $\{U_i, \varphi_i\}$ of (consistently oriented) coordinate charts, then define the integral

$$\int_{\Omega} \alpha \equiv \sum_i \int_{U_i} \psi_i \alpha,$$

where each term in the sum is evaluated by pulling back to R^n as described above. This quantity is well-defined; that is, it does not depend on the choice of the coordinate charts, nor the partition of unity.

Stokes' theorem reads: If ω is an $(n-1)$-form with compact support on Ω and $-\Omega$ denotes the boundary of Ω with its induced orientation, then

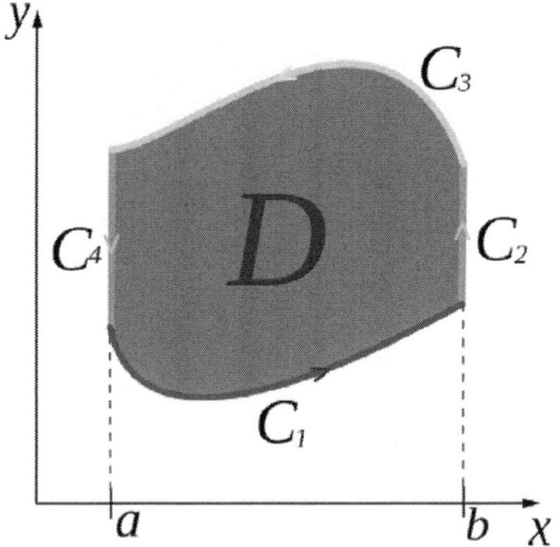

Figure: *A "normal" integration manifold (here called* D *instead of* Ω*) for the special case* n = 2

$$\int_{\Omega} \mathrm{d}\omega = \int_{\partial\Omega} \omega \left(= \oint_{\partial\Omega} \omega \right).$$

Here d is the exterior derivative, which is defined using the manifold structure only. On the r.h.s., a circle is sometimes used within the integral sign to stress the fact that the $(n-1)$-manifold "Ω is *closed.*[5] The r.h.s. of the equation is often used to formulate *integral* laws; the l.h.s. then leads to equivalent *differential* formulations.

The theorem is often used in situations where Ω is an embedded oriented submanifold of some bigger manifold on which the form Ω is defined.

A proof becomes particularly simple if the submanifold Ω is a so-called "normal manifold", as in the figure on the r.h.s., which can be segmented into vertical stripes (e.g. parallel to the x_n direction), such that after a partial integration concerning this variable, nontrivial contributions come only from the upper and lower boundary surfaces (coloured in yellow and red, respectively), where the complementary mutual orientations are visible through the arrows.

Topological Preliminaries; Integration over Chains

Let *M* be a smooth manifold. A smooth singular k-simplex of *M* is a smooth map from the standard simplex in R^k to *M*. The free abelian group, S_k, generated by singular *k*-simplices is said to consist of singular *k*-chains of *M*. These groups, together with boundary map, ", define a chain complex. The corresponding homology (resp. cohomology) is called the smooth singular homology (resp. cohomology) of *M*.

On the other hand, the differential forms, with exterior derivative, *d*, as the connecting map, form a cochain complex, which defines de Rham cohomology.

Differential *k*-forms can be integrated over a *k*-simplex in a natural way, by pulling back to R^k. Extending by linearity allows one to integrate over chains. This gives a linear map from the space of *k*-forms to the *k*-th group in the singular cochain, S_k^*, the linear functionals on S_k. In other words, a *k*-form ω defines a functional

$$I(\omega)(c) = \oint_c \omega$$

on the *k*-chains. Stokes' theorem says that this is a chain map from de Rham cohomology to singular cohomology; the exterior derivative, *d*, behaves like the *dual* of " on forms. This gives a homomorphism from de Rham cohomology to singular cohomology. On the level of forms, this means:

1. closed forms, i.e., $d\omega = 0$, have zero integral over *boundaries*, i.e. over manifolds that can be written as $\partial \sum_c M_c$, and
2. exact forms, i.e., $\omega = d\sigma$, have zero integral over *cycles*, i.e. if the boundaries sum up to the empty set: $\sum_c \partial M_c = \varnothing$.

De Rham's theorem shows that this homomorphism is in fact an isomorphism. So the converse to 1 and 2 above hold true. In other words, if $\{c_i\}$ are cycles generating the *k*-th homology group, then for any corresponding real numbers, $\{a_i\}$, there exist a closed form, ω, such that

$$\oint_{c_i} \omega = a_i,$$

and this form is unique up to exact forms.

Underlying Principle

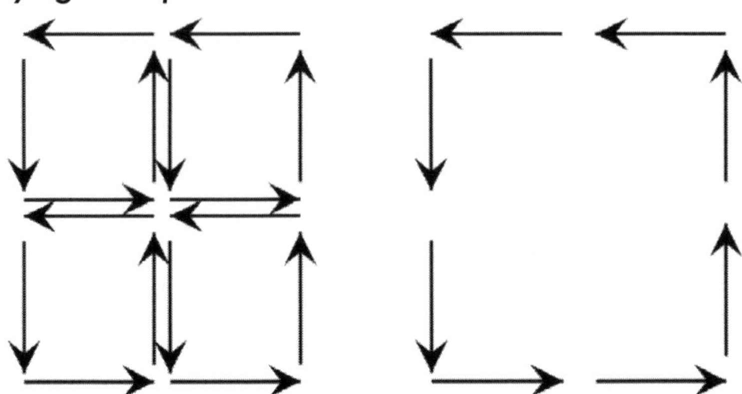

To simplify these topological arguments, it is worthwhile to examine the underlying principle by considering an example for $d = 2$ dimensions. The essential idea can be understood by the diagram on the left, which shows that, in an oriented tiling of a manifold, the interior paths are traversed in opposite directions; their contributions to the path integral thus cancel each other pairwise. As a consequence, only the contribution from the boundary remains. It thus suffices to prove Stokes' theorem for sufficiently fine tilings (or, equivalently, simplices), which usually is not difficult.

Special Cases

The general form of the Stokes theorem using differential forms is more powerful and easier to use than the special cases. The traditional versions can be formulated using Cartesian coordinates without the machinery of differential geometry, and thus are more accessible. Further, they are older and their names are more familiar as a result. The traditional forms are often considered more convenient by practicing scientists and engineers but the non-naturalness of the traditional formulation becomes apparent when using other coordinate systems, even familiar ones like spherical or cylindrical coordinates. There is potential for confusion in the way names are applied, and the use of dual formulations.

Kelvin–Stokes Theorem

This is a (dualized) 1+1 dimensional case, for a 1-form (dualized because it is a statement about vector fields). This special case is often just referred to as the *Stokes' theorem* in many introductory university vector calculus courses and as used in physics and engineering. It is also sometimes known as the curl theorem.

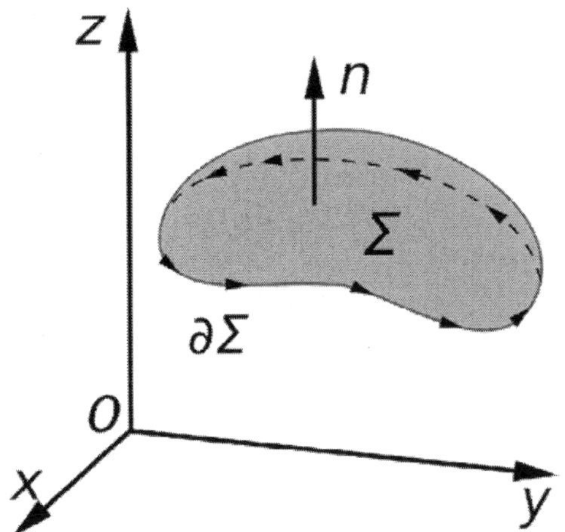

Figure: *An illustration of the Kelvin–Stokes theorem, with surface* Σ*, its boundary* $\partial\Sigma$ *and the "normal" vector* n.

The classical Kelvin–Stokes theorem:

$$\int_{\Sigma} \nabla \times \mathbf{F} \cdot d\mathbf{\Sigma} = \oint_{\partial\Sigma} \mathbf{F} \cdot d\mathbf{r},$$

which relates the surface integral of the curl of a vector field over a surface Σ in Euclidean three-space to the line integral of the vector field over its boundary, is a special case of the general Stokes theorem (with $n = 2$) once we identify a vector field with a 1 form using the metric on Euclidean three-space. The curve of the line integral, “Σ, must have positive orientation, meaning that dr points counterclockwise when the surface normal, $d\Sigma$, points toward the viewer, following the right-hand rule.

One consequence of the Kelvin–Stokes theorem is that the field lines of a vector field with zero curl cannot be closed contours. The formula can be rewritten as:

$$\iint_{\Sigma} \left\{ \left(\frac{\partial R}{\partial y} - \frac{\partial Q}{\partial z} \right) dy dz + \left(\frac{\partial P}{\partial z} - \frac{\partial R}{\partial x} \right) dz dx + \left(\frac{\partial Q}{\partial x} - \frac{\partial P}{\partial y} \right) dx dy \right\}$$

$$= \oint_{\partial\Sigma} \{ P dx + Q dy + R dz \},$$

where P, Q and R are the components of F.

These variants are rarely used:

$$\int_{\Sigma} \left[g(\nabla \times \mathbf{F}) + (\nabla g) \times \mathbf{F} \right] \cdot d\mathbf{\Sigma} = \oint_{\partial\Sigma} g\mathbf{F} \cdot d\mathbf{r},$$

$$\int_{\Sigma} \left[\mathbf{F}(\nabla \cdot \mathbf{G}) - \mathbf{G}(\nabla \cdot \mathbf{F}) + (\mathbf{G} \cdot \nabla)\mathbf{F} - (\mathbf{F} \cdot \nabla)\mathbf{G} \right] \cdot d\mathbf{\Sigma} = \oint_{\partial\Sigma} (\mathbf{F} \times \mathbf{G}) \cdot d\mathbf{r}.$$

$$\int_{\Sigma} \nabla(\mathbf{F}\cdot d\mathbf{\Sigma}) - (\nabla\cdot\mathbf{F})d\mathbf{\Sigma} = \oint_{\partial\Sigma} d\mathbf{r}\times\mathbf{F}.$$

Green's Theorem

Green's theorem is immediately recognisable as the third integrand of both sides in the integral in terms of P, Q, and R cited above.

In Electromagnetism

Two of the four Maxwell equations involve curls of 3-D vector fields and their differential and integral forms are related by the Kelvin–Stokes theorem. Caution must be taken to avoid cases with moving boundaries: the partial time derivatives are intended to exclude such cases. If moving boundaries are included, interchange of integration and differentiation introduces terms related to boundary motion not included in the results below:

Name	*Differential form*	*Integral form (using Kelvin–Stokes theorem plus relativistic invariance, $\int \frac{\partial}{\partial t}\ldots \to \frac{d}{dt}\int\ldots$)*
Maxwell–Faraday equation Faraday's law of induction:	$\nabla\times\mathbf{E} = -\frac{\partial\mathbf{B}}{\partial t}$	$\oint_C \mathbf{E}\cdot d\mathbf{l} = \iint_S \nabla\times\mathbf{E}\cdot d\mathbf{A} = -\iint_S \frac{\partial\mathbf{B}}{\partial t}\cdot d\mathbf{A}$ (with C and S not necessarily stationary)
Ampère's law (with Maxwell's extension):	$\nabla\times\mathbf{H} = \mathbf{J} + \frac{\partial\mathbf{D}}{\partial t}$	$\oint_C \mathbf{H}\cdot d\mathbf{l} = \iint_S \nabla\times\mathbf{H}\cdot d\mathbf{A} = \iint_S \mathbf{J}\cdot d\mathbf{A} + \iint_S \frac{\partial\mathbf{D}}{\partial t}\cdot d\mathbf{A}$ (with C and S not necessarily stationary)

The above listed subset of Maxwell's equations are valid for electromagnetic fields expressed in SI units. In other systems of units, such as CGS or Gaussian units, the scaling factors for the terms differ. For example, in Gaussian units, Faraday's law of induction and Ampθre's law take the forms[6][7]

$$\nabla\times\mathbf{E} = -\frac{1}{c}\frac{\partial\mathbf{B}}{\partial t},$$

$$\nabla\times\mathbf{H} = \frac{1}{c}\frac{\partial\mathbf{D}}{\partial t} + \frac{4\pi}{c}\mathbf{J},$$

respectively, where c is the speed of light in vacuum.

Divergence Theorem

Likewise, the Ostrogradsky–Gauss theorem (also known as the divergence theorem or Gauss's theorem)

$$\int_{\mathrm{Vol}} \nabla\cdot\mathbf{F}\, d_{\mathrm{Vol}} = \oint_{\partial\mathrm{Vol}} \mathbf{F}\cdot d\mathbf{\Sigma}$$

is a special case if we identify a vector field with the $n-1$ form obtained by contracting the vector field with the Euclidean volume form.

Unit Vector System

In mathematics, a unit vector in a normed vector space is a vector (often a spatial vector) whose length is 1 (the unit length). A unit vector is often denoted by a lowercase letter with a "hat", like so: t (pronounced "i-hat").

The normalized vector or versor u of a non-zero vector u is the unit vector codirectional with u, i.e., where $\|\mathbf{u}\|$ is the norm (or length) of u. The term *normalized vector* is sometimes used as a synonym for *unit vector*.

The elements of a basis are usually chosen to be unit vectors. Every vector in the space may be written as a linear combination of unit vectors. The most commonly encountered bases are Cartesian, polar, and spherical coordinates. Each uses different unit vectors according to the symmetry of the coordinate system. Since these systems are encountered in so many different contexts, it is not uncommon to encounter different naming conventions than those used here.

By definition, in Euclidean space the dot product of two unit vectors is simply the cosine of the angle between them. In three-dimensional Euclidean space, the cross product of two orthogonal unit vectors is another unit vector, orthogonal to both of them.

Chapter 4

Introduction to Coordinate Geometry

Coordinate Geometry is a system of geometry where the position of points on the plane is described using an ordered pair of numbers.

Recall that a plane is a flat surface that goes on forever in both directions. If we were to place a point on the plane, coordinate geometry gives us a way to describe exactly where it is by using two numbers.

What are Coordinates?

	A	B	C	D	E	F
1						
2						
3				X		
4						
5						
6						

To introduce the idea, consider the grid on the right. The columns of the grid are lettered A,B,C etc. The rows are numbered 1,2,3 etc from the top. We can see that the X is in box D3; that is, column D, row 3.

D and 3 are called the *coordinates* of the box. It has two parts: the row and the column. There are many boxes in each row and many boxes in each column. But by having both we can find one single box, where the row and column intersect.

Coordinate Geometry

Coordinate geometry is one of the most important and exciting ideas of mathematics. In particular it is central to the mathematics students meet at school. It provides a connection between algebra and geometry through graphs of lines and curves. This enables geometric problems to be solved algebraically and provides geometric insights into algebra.

The invention of calculus was an extremely important development in mathematics that enabled mathematicians and physicists to model the real world in ways that was previously impossible. It brought together nearly all of algebra and geometry using the coordinate plane. The invention of calculus depended on the development of coordinate geometry.

It is expected that students have met plotting points on the plane and have plotted points from tables of values of both linear and non linear functions.

The number plane (Cartesian plane) is divided into four quadrants by two perpendicular axes called the x-axis (horizontal line) and the y-axis (vertical line). These axes intersect at a point called the origin. The position of any point in the plane can be represented by an ordered pair of numbers (x, y). These ordered pairs are called the coordinates of the point.

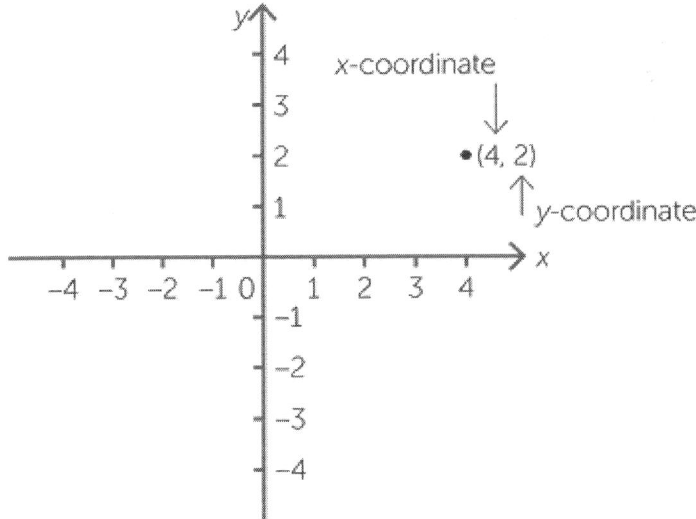

Figure: *The point with coordinates (4, 2) has been plotted on the Cartesian plane shown. The coordinates of the origin are (0, 0).*

Once the coordinates of two points are known the distance between the two points and midpoint of the interval joining the points can be found.

Distance between Two Points

Distances are always positive, or zero if the points coincide. The distance from A to B is the same as the distance from B to A.

The example above considered the special cases when the line interval AB is either horizontal or vertical. Pythagoras' theorem is used to calculate the distance between two points when the line interval between them is neither vertical nor horizontal.

The distance between the points A(1, 2) and B(4, 6) is calculated below.

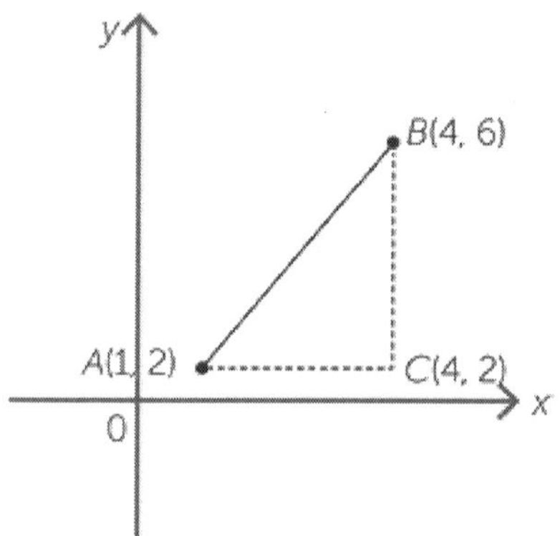

$$AC = 4 - 1 = 3 \text{ and } BC = 6 - 2 = 4.$$

By Pythagoras' theorem,

$$AB^2 = 3^2 + 4^2 = 25$$

And so AB = 5

The General Case

We can obtain a formula for the length of any interval. Suppose that $P(x_1, y_1)$ and $Q(x_2, y_2)$ are two points.

Form the right-angled triangle PQX, where X is the point (x_2, y_1),

$$PX = x_2 - x_1 \text{ or } x_1 - x_2 \text{ and } QX = y_2 - y_1 \text{ or } y_1 - y_2$$

depending on the positions of P and Q.

By Pythagoras' theorem:

$$PQ2 = PX_2 + QX_2$$
$$= (x_2 - x_1)_2 + (y_2 - y_1)^2$$

Therefore $PQ = QP = \sqrt{(X_2 - X_1)^2 + (y_2 - y_1)^2}$

Note that $(x_2 - x_1)^2$ is the same as $(x_1 - x_1)^2$ and therefore it doesn't matter whether we go from P to Q or from Q to P – the result is the same.

Gradient of an Interval

The gradient is a measure of the steepness of line. There are several ways to measure steepness. In coordinate geometry the standard way to define the gradient of an interval AB is $\frac{\text{Rise}}{\text{run}}$ where rise is the change in the y-values as you move from A to B and run is the change in the x-values as you move from A to B. We will usually the pronumeral m for gradient.

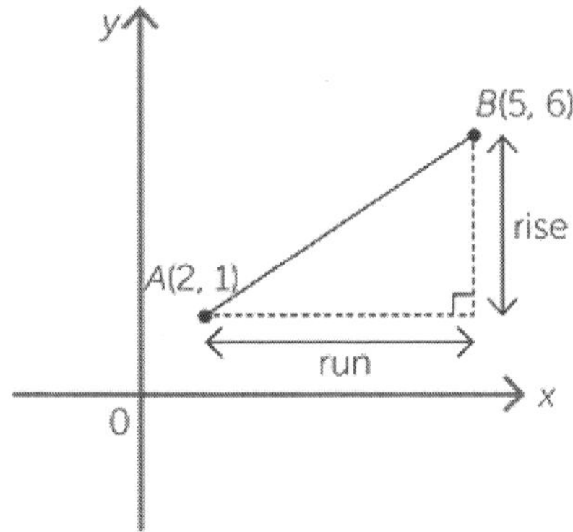

Given the points A(2, 1) and B(5, 6):

$$\text{gradient of interval AB} = \frac{\text{Rise}}{\text{run}} = \frac{6-1}{5-2} = \frac{5}{3}$$

Notice that as you move from A to B along the interval the y-value increases as the x-value increases. The gradient is positive.

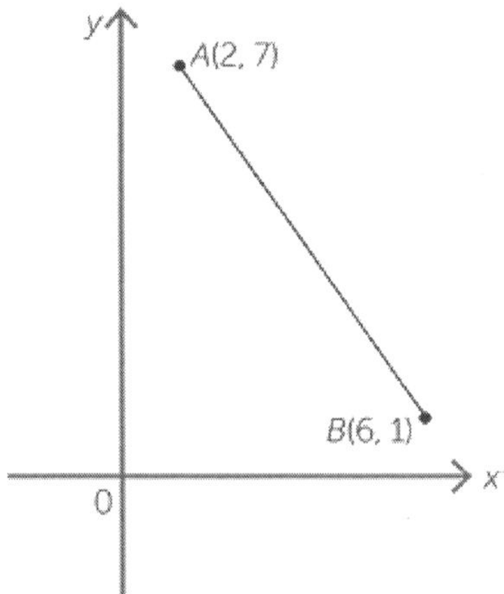

Given the points A(2, 7) and B(6, 1)

$$\text{gradient of interval AB} = \frac{\text{Rise}}{\text{run}} = \frac{1-7}{6-2} = \frac{-6}{4} = -\frac{3}{2}$$

or

$$\text{gradient of interval BA} = \frac{\text{Rise}}{\text{run}} = \frac{7-1}{2-6} = -\frac{3}{2}$$

Notice that in this case as we move from A to B the y value decreases as the x value increases. The gradient is negative. Similarly the gradient of BA $= -\frac{3}{2}$ which is the same as the gradient of AB.

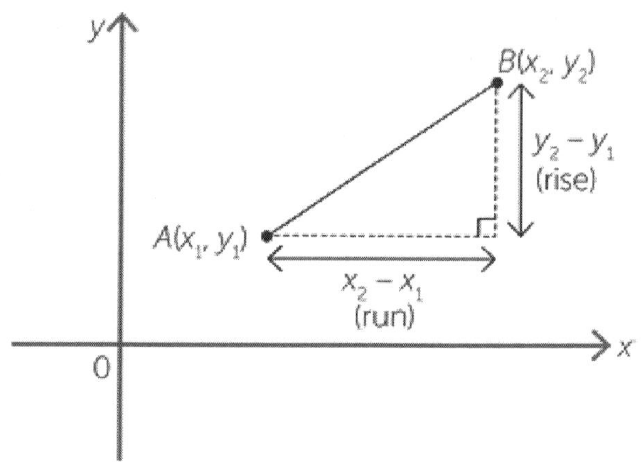

In general:

$$\text{gradient of line interval AB} = \frac{\text{rise}}{\text{run}}$$

$$m = \frac{y_2 - y_1}{x_2 - x_1}$$

Note that since $\frac{y_2 - y_1}{x_2 - x_1} = \frac{y_2 - y_1}{x_2 - x_1}$ it does not matter which point we take as the first and which point we take as the second.

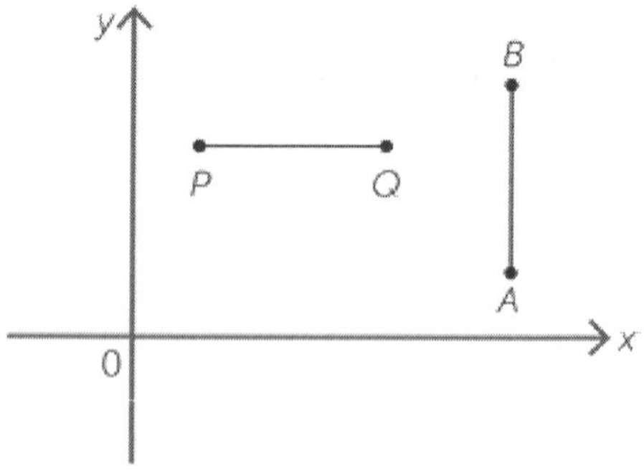

Gradient of *PQ* is zero
Gradient of *AB* is not defined

If the interval is vertical, the run is zero and the gradient of the interval is not defined. This is shown by interval AB.

If the interval is horizontal, the rise is zero as shown by interval PQ. The gradient of the interval is zero.

Gradient of a Line

The gradient of a line is defined to be the gradient of any interval within the line.

This definition depends on the fact that two intervals on a line have the same gradient.

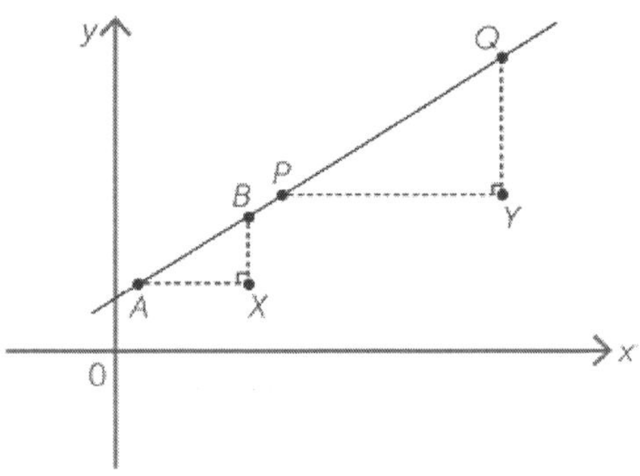

Suppose AB and PQ are two intervals on the same straight line. Draw right-angled triangles ABX and PQY with sides AX and PY parallel to the x-axis and sides BX and QY parallel to the y-axis.

Triangle ABX is similar to triangle PQY since the corresponding angles are equal. Therefore:

$$\frac{QY}{PY} - \frac{BX}{AX}$$

That is, the intervals have the same gradient.

Intercepts

The x-intercept of a line is the point at which it crosses the x-axis.

The y-intercept of a line is the point at which it crosses the y-axis.

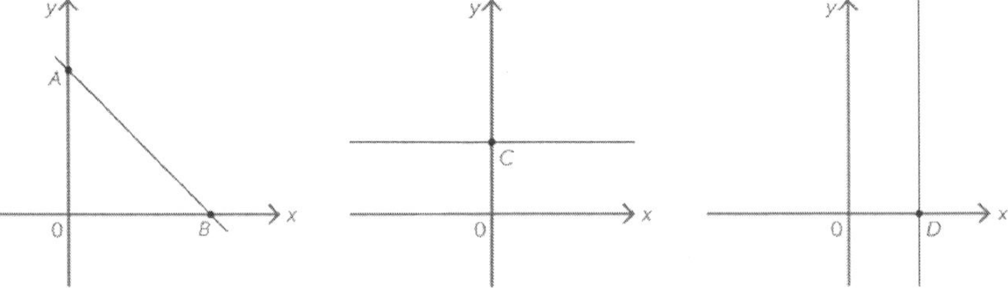

In the diagram to the left the y-intercept is at A and the x-intercept at B.

The second diagram shows a line parallel to the x-axis and it has a y-intercept at C.

The third diagram shows a line parallel to the y-axis and it has an x-intercept at D.

Equation of a straight line

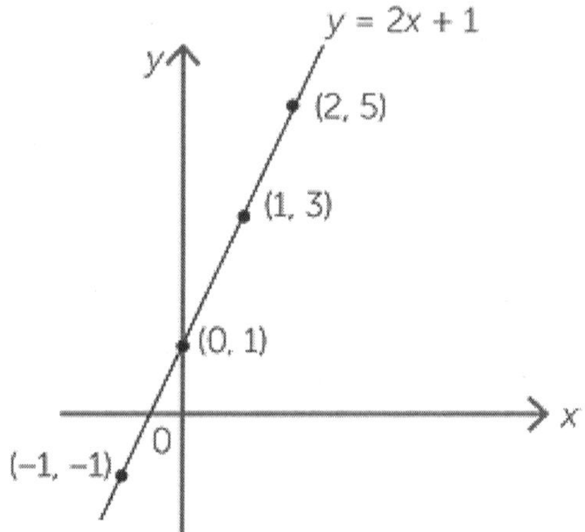

When we plot points which satisfy the equation y = 2x + 1 we find that they lie in a straight line.

Can we find the equation of the line given suitable geometric information about the line? The following shows that this can be done given the gradient of the line and the y-intercept.

The line y = 3x + 2

Consider the line with gradient 3 and y-intercept 2. This passes through the point A(0, 2). Let B(x, y) be any point on this line.

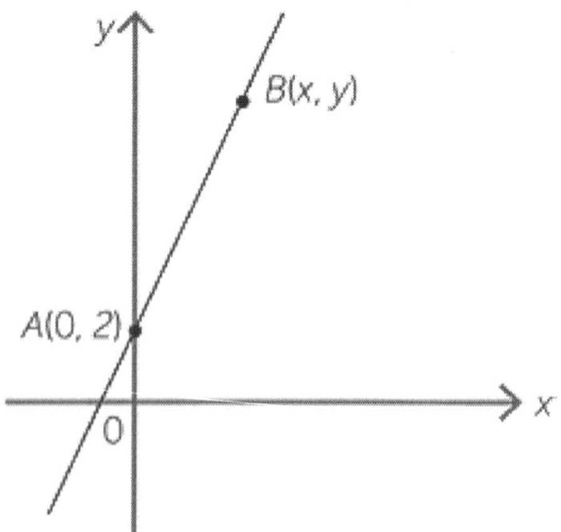

$$\text{Gradient of interval AB} = \frac{rise}{run} = \frac{y-2}{x-0} = \frac{y-2}{x}$$

The gradient of the line is 3.

So, $$\frac{y-2}{x} = 3$$

Rearranging $$Y - 2 = 3x$$ $$Y = 3x + 2$$

So the coordinates of B(x, y) satisfies y = 3x + 2. This is called the equation of the line.

Conversely suppose that B(x, y) satisfies the equation y = 3x + 2, then $\frac{y-2}{x-0} = 3$ and it passes the point (0, 2) so the point lies on the line with gradient 3 and y-intercept 2.

We summarise this by saying that the equation of the line is y = 3x + 2.

The equation y = mx + c

Consider the line with gradient m and y-intercept c. If passes through the point A(0, c).

Let B(x, y) be any point on this line.

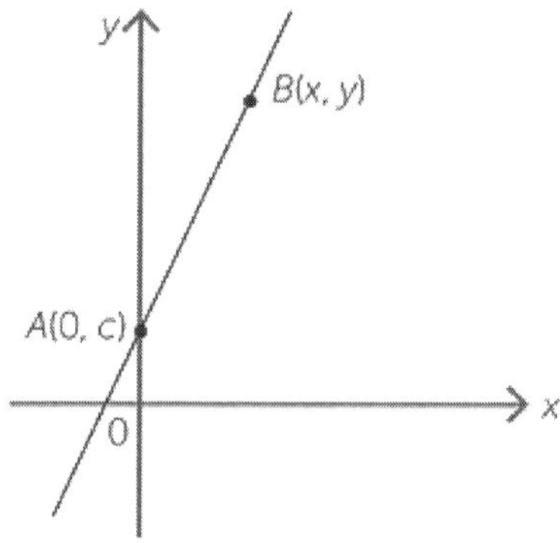

Gradient of interval $$AB = \frac{y-c}{x-0} = \frac{y-c}{x}$$

We know the gradient of the line is m.

$$\frac{y-c}{x}=m$$

Therefore $\quad y-c=mx$

$$y=mx+c$$

That is, the line in the cartesian plane with gradient m and y-intercept c has equation y = mx + c. Conversely, the points whose coordinates satisfy the equation y = mx + c always lie on the line with gradient m and y-intercept c.

Vertical and Horizontal Lines

Vertical Lines

In a vertical line all points have the same x-coordinate, but the y-coordinate can take any value. The equation of the vertical line through the point (6, 0) is x = 6. The x-axis intercept is 6. All the points on this line have x-coordinate 6.

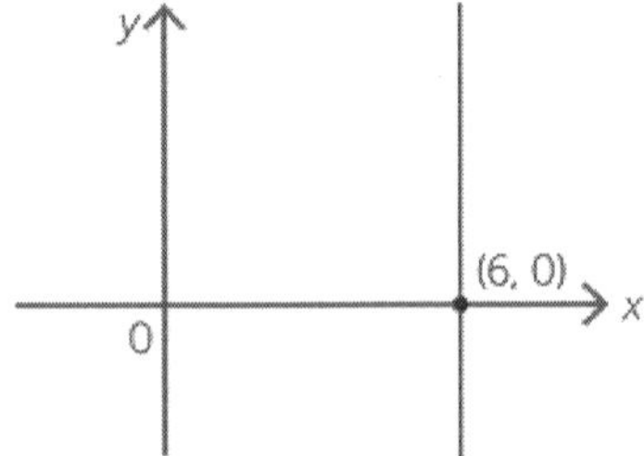

In general, the equation of the vertical line through P(a, b) is x = a. Because this line does not have a gradient it cannot be written in the form

y = mx + b.

Horizontal Lines

A horizontal line has gradient 0. In a horizontal line all points have the same y-coordinate, but the x-coordinate can take any value. The equation of the horizontal line through the point (0, 5) is y = 5. The equation of the horizontal line through the point (9, 5) is y = 5.

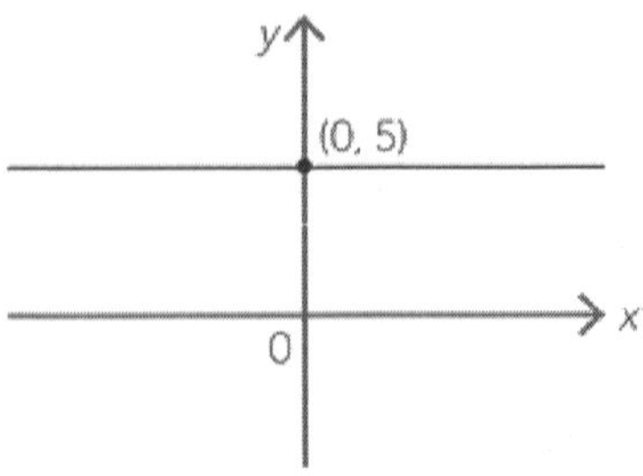

In general, the equation of the horizontal line through P(a, b) is y = b.

The Coordinate Plane

In coordinate geometry, points are placed on the "coordinate plane" as shown below. It has two scales - one running across the plane called the "x axis" and another a right angles to it called the y axis. The point where the axes cross is called the origin and is where both x and y are zero.

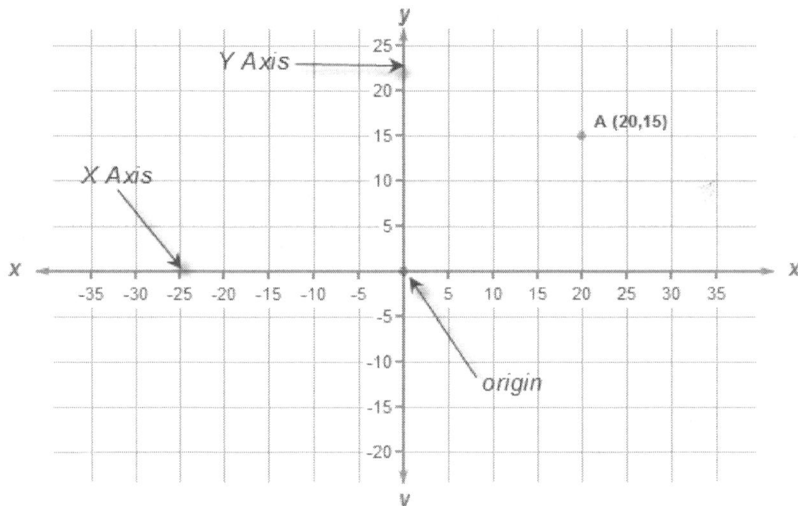

On the x-axis, values to the right are positive and those to the left are negative.

On the y-axis, values above the origin are positive and those below are negative.

A point's location on the plane is given by two numbers, the first tells where it is on the x-axis and the second which tells where it is on the y-axis. Together, they define a single, unique position on the plane. So in the diagram above, the point A has an x value of 20 and a y value of 15. These are the coordinates of the point A, sometimes referred to as its "rectangular coordinates". Note that the order is important; the x coordinate is always the first one of the pair.

As you remember from pre-algebra a coordinate plane is a two-dimensional number line where the vertical line is called the y-axis and the horizontal is called the x-axis. These lines are perpendicular and intersect at their zero points. This point is called the origin. The axes divide the plane into four quadrants.

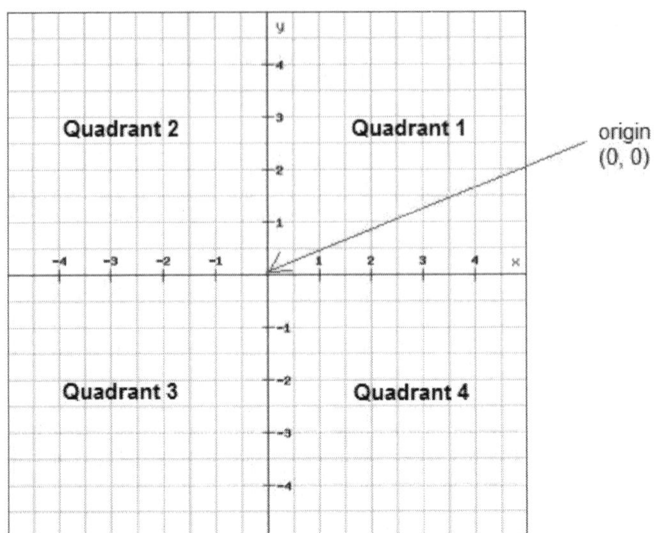

A point in a coordinate plane is named by its ordered pair of the form of (x, y). The first number corresponds to the x-coordinates and the second to the y-coordinate.

The completeness property for points in the plane tells us two things

1. Exactly one point in the plane is named given the numbers of the ordered pair and
2. Exactly one ordered pair of numbers names a given point in the plane.

To graph a point one draws a dot at the coordinates that corresponds to the ordered pair. It's always a god idea to start at the origin. The x-coordinate tells you have many steps you have to take to the right (positive) or left (negative). And the y-coordinate tells you have many steps to move up (positive) or down (negative).

To find out the coordinates of a point in the coordinate system you do the opposite. Begin at the point and follow a vertical line either up or down to the x-axis. There is your x-coordinate. And then do the same but following a horizontal line to find the y-coordinate. A relation is a set of ordered pairs. The first coordinate (usually the x-coordinate) is called the domain and the second (usually the y-coordinate) is called the range.

Videolesson: Draw the following ordered pairs in the coordinate plane.

(0, 0); (0,4); (4, -2); (-2, -4); (1, 3)

Coordinates of a Point

Every point in space can be assigned three numbers with respect to a starting point. Those three numbers allow us to distinguish any point from any other point in space.

- *Coordinates of a point:* Each point on a number line is assigned a number. In the same way, each point in a plane is assigned a pair of numbers.
- *x axis and y axis:* To locate points in a plane, two perpendicular lines are used—a horizontal line called the x axis and a vertical line called the y axis.
- *Origin:* The point of intersection of the x axis and y axis.
- *Coordinate plane:* The x axis, y axis, and all the points in the plane they determine.
- *Ordered pairs:* Every point in a coordinate plane is named by a pair of numbers whose order is important; these numbers are written in parentheses and separated by a comma.
- *x coordinate:* The number to the left of the comma in an ordered pair is the x coordinate of the point and indicates the amount of movement along the x axis from the origin. The movement is to the right if the number is positive and to the left if the number is negative.
- *y coordinate:* The number to the right of the comma in an ordered pair is the y coordinate of the point and indicates the amount of movement perpendicular to the x axis. The movement is above the x axis if the number is positive and below the x axis if the number is negative.

Note: The coordinates [ordered pair] for the origin are (0, 0).

The x axis and y axis separate the coordinate plane into four regions called quadrants. The upper right quadrant is quadrant 1; the upper left quadrant is quadrant II; the lower left quadrant is quadrant III; and the lower right quadrant is quadrant IV. Notice the following:

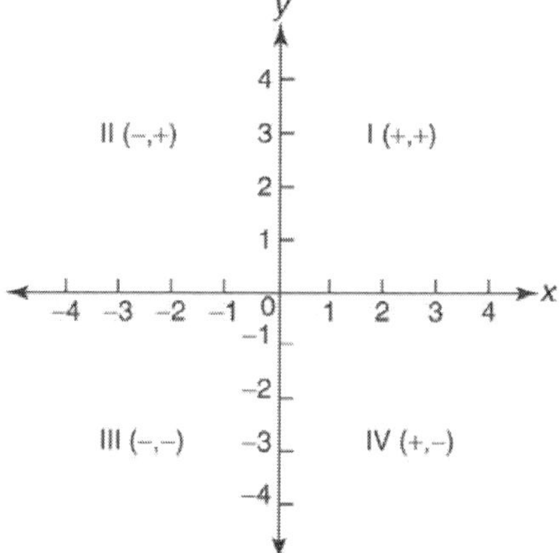

Figure: *The coordinate axes separate the plane into four quadrants.*

- In quadrant I, x is always positive and y is always positive.
- In quadrant II, x is always negative and y is always positive.
- In quadrant III, x is always negative and y is always negative.
- In quadrant IV, x is always positive and y is always negative.

The point associated with an ordered pair of real numbers is called the graph of the ordered pair.

Example 1: Identify the points *A, B, C, D, E,* and *F* on the coordinate graph in Figure .

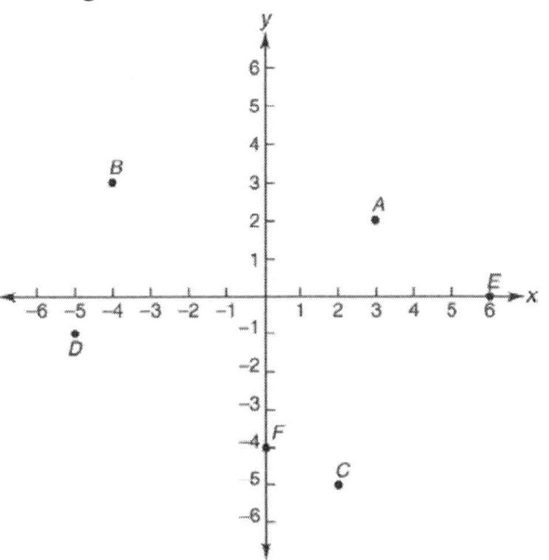

Figure: *Finding the coordinates of specific points in the plane.*

Example 2: Rectangle *ABCD* has coordinates as follows: *A*(–5,2), *B*(8,2), and *C*(8, –4). Find the coordinates of *D*.

A graph is helpful in solving this problem. Refer to Figure . The coordinates of *D* must be (–5,–4) .

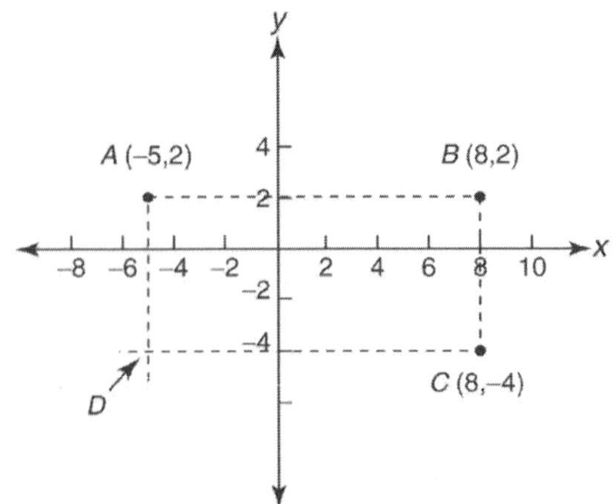

Figure: *Finding the coordinates of the fourth vertex of a rectangle.*

Plane

In mathematics, a plane is a flat, two-dimensional surface. A plane is the two-dimensional analogue of a point (zero dimensions), a line (one dimension) and three-dimensional space. Planes can arise as subspaces of some higher-dimensional space, as with the walls of a room, or they may enjoy an independent existence in their own right, as in the setting of Euclidean geometry.

When working exclusively in two-dimensional Euclidean space, the definite article is used, so, *the* plane refers to the whole space. Many fundamental tasks in mathematics, geometry, trigonometry, graph theory and graphing are performed in a two-dimensional space, or in other words, in the plane.

Euclid set forth the first great landmark of mathematical thought, an axiomatic treatment of geometry.[1] He selected a small core of undefined terms (called *common notions*) and postulates (or axioms) which he then used to prove various geometrical statements. Although the plane in its modern sense is not directly given a definition anywhere in the *Elements*, it may be thought of as part of the common notions.[2] In his work Euclid never makes use of numbers to measure length, angle, or area. In this way the Euclidean plane is not quite the same as the Cartesian plane.

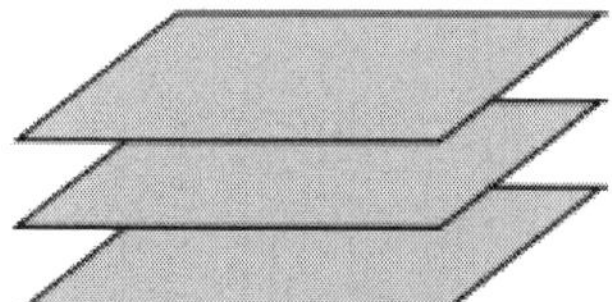

Figure: *Three parallel planes.*

Planes Embedded in 3-Dimensional Euclidean Space

This section is solely concerned with planes embedded in three dimensions: specifically, in R3.

Properties

The following statements hold in three-dimensional Euclidean space but not in higher dimensions, though they have higher-dimensional analogues:

- Two distinct planes are either parallel or they intersect in a line.
- A line is either parallel to a plane, intersects it at a single point, or is contained in the plane.

- Two distinct lines perpendicular to the same plane must be parallel to each other.
- Two distinct planes perpendicular to the same line must be parallel to each other.

Point-Normal Form and General Form of the Equation of a Plane

In a manner analogous to the way lines in a two-dimensional space are described using a point-slope form for their equations, planes in a three dimensional space have a natural description using a point in the plane and a vector (the normal vector) to indicate its "inclination".

Specifically, let $\mathbf{r}_0$ be the position vector of some point $P_0 = (x_0, y_0, z_0)$, and let $\mathbf{n} = (a, b, c)$ be a nonzero vector. The plane determined by this point and vector consists of those points *P*, with position vector r, such that the vector drawn from P_0 to P is perpendicular to *n*. Recalling that two vectors are perpendicular if and only if their dot product is zero, it follows that the desired plane can be described as the set of all points r such that

$$\mathbf{n} \cdot (\mathbf{r} - \mathbf{r}_0) = 0.$$

(The dot here means a dot product, not scalar multiplication.) Expanded this becomes

$$a(x - x_0) + b(y - y_0) + c(z - z_0) = 0,$$

which is the *point-normal* form of the equation of a plane. This is just a linear equation:

$$ax + by + cz + d = 0, \text{ where } d = -(ax_0 + by_0 + cz_0).$$

Conversely, it is easily shown that if *a*, *b*, *c* and *d* are constants and *a*, *b*, and *c* are not all zero, then the graph of the equation

$$ax + by + cz + d = 0,$$

is a plane having the vector $\mathbf{n} = (a, b, c)$ as a normal. This familiar equation for a plane is called the *general form* of the equation of the plane.

Thus for example a regression equation of the form $y = d + ax + cz$ (with *b*=-1) establishes a best-fit plane in three-dimensional space when there are two explanatory variables.

Describing a Plane with a Point and Two Vectors Lying on it

Alternatively, a plane may be described parametrically as the set of all points of the form

$$\mathbf{r} = \mathbf{r}_0 + s\mathbf{v} + t\mathbf{w},$$

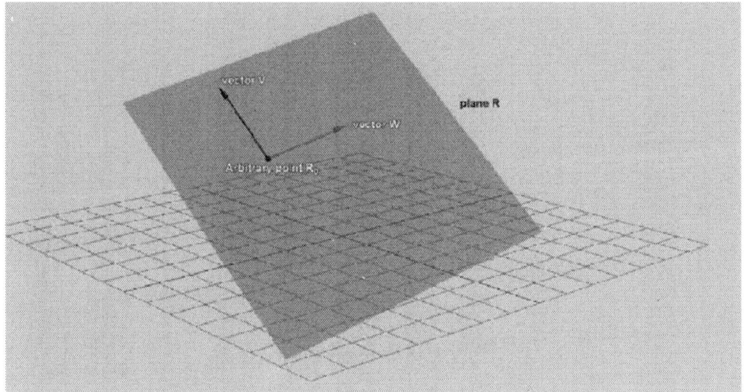

Figure: *Vector description of a plane*

where s and t range over all real numbers, v and w are given linearly independent vectors defining the plane, and r_0 is the vector representing the position of an arbitrary (but fixed) point on the plane. The vectors v and w can be visualized as vectors starting at r_0 and pointing in different directions along the plane. Note that v and w can be perpendicular, but cannot be parallel.

Describing a Plane through Three Points

Let $p_1=(x_1, y_1, z_1)$, $p_2=(x_2, y_2, z_2)$, and $p_3=(x_3, y_3, z_3)$ be non-collinear points.

Method 1

The plane passing through p_1, p_2, and p_3 can be described as the set of all points (x,y,z) that satisfy the following determinant equations:

$$\begin{vmatrix} x-x_1 & y-y_1 & z-z_1 \\ x_2-x_1 & y_2-y_1 & z_2-z_1 \\ x_3-x_1 & y_3-y_1 & z_3-z_1 \end{vmatrix} = \begin{vmatrix} x-x_1 & y-y_1 & z-z_1 \\ x-x_2 & y-y_2 & z-z_2 \\ x-x_3 & y-y_3 & z-z_3 \end{vmatrix} = 0.$$

Method 2

To describe the plane by an equation of the form $ax+by+cz+d=0$, solve the following system of equations:

$$ax_1+by_1+cz_1+d=0$$

$$ax_2+by_2+cz_2+d=0$$

$$ax_3+by_3+cz_3+d=0.$$

This system can be solved using Cramer's Rule and basic matrix manipulations. Let

$$D = \begin{vmatrix} x_1 & y_1 & z_1 \\ x_2 & y_2 & z_2 \\ x_3 & y_3 & z_3 \end{vmatrix}.$$

If D is non-zero (so for planes not through the origin) the values for a, b and c can be calculated as follows:

$$a = \frac{-d}{D}\begin{vmatrix} 1 & y_1 & z_1 \\ 1 & y_2 & z_2 \\ 1 & y_3 & z_3 \end{vmatrix}$$

$$b = \frac{-d}{D}\begin{vmatrix} x_1 & 1 & z_1 \\ x_2 & 1 & z_2 \\ x_3 & 1 & z_3 \end{vmatrix}$$

$$c = \frac{-d}{D}\begin{vmatrix} x_1 & y_1 & 1 \\ x_2 & y_2 & 1 \\ x_3 & y_3 & 1 \end{vmatrix}.$$

These equations are parametric in d. Setting d equal to any non-zero number and substituting it into these equations will yield one solution set.

Method 3

This plane can also be described by the "point and a normal vector" prescription above. A suitable normal vector is given by the cross product

$$\mathbf{n} = (\mathbf{p}_2 - \mathbf{p}_1) \times (\mathbf{p}_3 - \mathbf{p}_1),$$

and the point r_0 can be taken to be any of the given points p_1,p_2 or p_3.

Distance from a Point to a Plane

In Euclidean space, the point on a plane $ax + by + cz = d$ that is closest to the origin has the Cartesian coordinates (x, y, z), where

$$x = \frac{ad}{a^2 + b^2 + c^2}, \qquad y = \frac{bd}{a^2 + b^2 + c^2}, \qquad z = \frac{cd}{a^2 + b^2 + c^2}.$$

From this the distance from the origin to the plane can be found. If what is desired is the distance from a point not at the origin to the nearest point on a plane, this can be found by a change of variables that moves the origin to coincide with the given point.

Converting General Problem to Distance-From-Origin Problem

Suppose we wish to find the nearest point on a plane to the point (X_0, Y_0), where the plane is given by $aX + bY + cZ = D$. We define $x = X - X_0$, $y = Y - Y_0$, $z = Z - Z_0$, and $d = D - aX_0 - bY_0 - cZ_0$, to obtain $ax + by + cz = d$ as the plane expressed in terms of the transformed variables. Now the problem has become one of finding the nearest point on this plane to the origin, and its distance from the origin. The point on the plane in terms of the original coordinates can be found from this point using the above relationships between x and X, between y and Y, and between z and Z; the distance in terms of the original coordinates is the same as the distance in terms of the revised coordinates.

Restatement Using Linear Algebra

The formula for the closest point to the origin may be expressed more succinctly using notation from linear algebra. The expression $ax + by + cz$ in the definition of a plane is a dot product $(a,b,c)\cdot(x,y,z)$, and the expression $a^2 + b^2 + c^2$ appearing in the solution is the squared norm $|(a,b,c)|^2$. Thus, if $\mathbf{v} = (a,b,c)$ is a given vector, the plane may be described as the set of vectors w for which $\mathbf{v}\cdot\mathbf{w} = d$ and the closest point on this plane is the vector

$$\mathbf{p} = \frac{\mathbf{v}d}{|\mathbf{v}|^2}.$$

The Euclidean distance from the origin to the plane is the norm of this point,

$$\frac{d}{|\mathbf{v}|} = \frac{d}{\sqrt{a^2 + b^2 + c^2}}.$$

Why this is the Closest Point

In either the coordinate or vector formulations, one may verify that the given point lies on the given plane by plugging the point into the equation of the plane.

To see that it is the closest point to the origin on the plane, observe that p is a scalar multiple of the vector v defining the plane, and is therefore orthogonal to the plane. Thus, if q is any point on the plane other than itself, then the line segments from the origin to $\mathbf{p}$ and from $\mathbf{p}$ to $\mathbf{q}$ form a right triangle, and by the Pythagorean theorem the distance from the origin to $\mathbf{q}$ is

$$\sqrt{|\mathbf{p}|^2 + |\mathbf{p} - \mathbf{q}|^2}$$

Since $|\mathbf{p}-\mathbf{q}|^2$ must be a positive number, this distance is greater than $|\mathbf{p}|$, the distance from the origin to $\mathbf{p}$.[2]

Alternatively, it is possible to rewrite the equation of the plane using dot products with $\mathbf{p}$ in place of the original dot product with v(because these two vectors are scalar multiples of each other) after which the fact that $\mathbf{p}$ is the closest point becomes an immediate consequence of the Cauchy–Schwarz inequality.

Line of Intersection between Two Planes

The line of intersection between two planes $\Pi_1 : \mathbf{n}_1 \cdot \mathbf{r} = h_1$ and $\Pi_2 : \mathbf{n}_2 \cdot \mathbf{r} = h_2$ where are normalized is given by

$$\mathbf{r} = (c_1\mathbf{n}_1 + c_2\mathbf{n}_2) + \lambda(\mathbf{n}_1 \times \mathbf{n}_2)$$

where

$$c_1 = \frac{h_1 - h_2(\mathbf{n}_1 \cdot \mathbf{n}_2)}{1 - (\mathbf{n}_1 \cdot \mathbf{n}_2)^2}$$

$$c_2 = \frac{h_2 - h_1(\mathbf{n}_1 \cdot \mathbf{n}_2)}{1 - (\mathbf{n}_1 \cdot \mathbf{n}_2)^2}.$$

This is found by noticing that the line must be perpendicular to both plane normals, and so parallel to their cross product $\mathbf{n}_1 \times \mathbf{n}_2$ (this cross product is zero if and only if the planes are parallel, and are therefore non-intersecting or entirely coincident).

The remainder of the expression is arrived at by finding an arbitrary point on the line. To do so, consider that any point in space may be written as $\mathbf{r} = c_1\mathbf{n}_1 + c_2\mathbf{n}_2 + \lambda(\mathbf{n}_1 \times \mathbf{n}_2)$, since $\{\mathbf{n}_1, \mathbf{n}_2, (\mathbf{n}_1 \times \mathbf{n}_2)\}$ is a basis. We wish to find a point which is on both planes (i.e. on their intersection), so insert this equation into each of the equations of the planes to get two simultaneous equations which can be solved for c_1 and c_2.

If we further assume that $\mathbf{n}_1$ and $\mathbf{n}_2$ are orthonormal then the closest point on the line of intersection to the origin is $\mathbf{r}_0 = h_1\mathbf{n}_1 + h_2\mathbf{n}_2$. If that is not the case, then a more complex procedure must be used.[8]

Dihedral Angle

Given two intersecting planes described by $\Pi_1 : a_1x + b_1y + c_1z + d_1 = 0$ and $\Pi_2 : a_2x + b_2y + c_2z + d_2 = 0$, the dihedral angle between them is defined to be the angle α between their normal directions:

$$\cos\alpha = \frac{\hat{n}_1 \cdot \hat{n}_2}{|\hat{n}_1||\hat{n}_2|} = \frac{a_1a_2 + b_1b_2 + c_1c_2}{\sqrt{a_1^2 + b_1^2 + c_1^2}\sqrt{a_2^2 + b_2^2 + c_2^2}}.$$

Planes in Various Areas of Mathematics

In addition to its familiar geometric structure, with isomorphisms that are isometries with respect to the usual inner product, the plane may be viewed at various other levels of abstraction. Each level of abstraction corresponds to a specific category.

At one extreme, all geometrical and metric concepts may be dropped to leave the topological plane, which may be thought of as an idealized homotopically trivial infinite rubber sheet, which retains a notion of proximity, but has no distances. The topological plane has a concept of a linear path, but no concept of a straight line. The topological plane, or its equivalent the open disc, is the basic topological neighbourhood used to construct surfaces (or 2-manifolds) classified in low-dimensional topology. Isomorphisms of the topological plane are all continuous bijections. The topological plane is the natural context for the branch of graph theory that deals with planar graphs, and results such as the four colour theorem.

The plane may also be viewed as an affine space, whose isomorphisms are combinations of translations and non-singular linear maps. From this viewpoint there are no distances, but collinearity and ratios of distances on any line are preserved.

Differential geometry views a plane as a 2-dimensional real manifold, a topological plane which is provided with a differential structure. Again in this case, there is no notion of distance, but there is now a concept of smoothness of maps, for example a differentiable or smooth path (depending on the type of differential structure applied). The isomorphisms in this case are bijections with the chosen degree of differentiability.

In the opposite direction of abstraction, we may apply a compatible field structure to the geometric plane, giving rise to the complex plane and the major area of complex analysis. The complex field has only two isomorphisms that leave the real line fixed, the identity and conjugation.

In the same way as in the real case, the plane may also be viewed as the simplest, one-dimensional (over the complex numbers) complex manifold, sometimes called the complex line. However, this viewpoint contrasts sharply with the case of the plane as a 2-dimensional real manifold. The isomorphisms are all conformal bijections of the complex plane, but the only possibilities are maps that correspond to the composition of a multiplication by a complex number and a translation.

In addition, the Euclidean geometry (which has zero curvature everywhere) is not the only geometry that the plane may have. The plane may be given a spherical geometry by using the stereographic projection. This can be thought of as placing a sphere on the plane (just like a ball on the floor), removing the top point, and projecting the sphere onto the plane from this point). This is one of the projections that may be used in making a flat map of part of the Earth's surface. The resulting geometry has constant positive curvature.

Alternatively, the plane can also be given a metric which gives it constant negative curvature giving the hyperbolic plane. The latter possibility finds an application in the theory of special relativity in the simplified case where there are two spatial dimensions and one time dimension. (The hyperbolic plane is a timelike hypersurface in three-dimensional Minkowski space.)

Topological and Differential Geometric Notions

The one-point compactification of the plane is homeomorphic to a sphere; the open disk is homeomorphic to a sphere with the "north pole" missing; adding that point completes the (compact) sphere. The result of this compactification is a manifold referred to as the Riemann sphere or the complex projective line. The projection from the Euclidean plane to a sphere without a point is a diffeomorphism and even a conformal map.

The plane itself is homeomorphic (and diffeomorphic) to an open disk. For the hyperbolic plane such diffeomorphism is conformal, but for the Euclidean plane it is not.

Distance from a Point to a Line

The distance (or perpendicular distance) from a point to a line is the shortest distance from a point to a line in Euclidean geometry. It is the length of the line segment which joins the point to the line and is perpendicular to the line. The formula for calculating it can be derived and expressed in several ways.

Knowing the shortest distance from a point to a line can be useful in various situations—for example, finding the shortest distance to reach a road, quantifying the scatter on a graph, etc. In Deming regression, a type of linear curve fitting, if the dependent and independent variables have equal variance this results in orthogonal regression in which the degree of imperfection of the fit is measured for each data point as the perpendicular distance of the point from the regression line.

Cartesian Coordinates

In the case of a line in the plane given by the equation $ax + by + c = 0$, where a, b and c are real constants with a and b not both zero, the distance from the line to a point (x_0,y_0) is

$$\text{distance}(ax+by+c=0,(x_0,y_0)) = \frac{|ax_0+by_0+c|}{\sqrt{a^2+b^2}}.$$

If the point on this line which is closest to (x_0,y_0) has coordinates (x_1,y_1), then:

$$x_1 = \frac{b(bx_0-ay_0)-ac}{a^2+b^2} \text{ and } y_1 = \frac{a(-bx_0+ay_0)-bc}{a^2+b^2}.$$

Vector Formulation

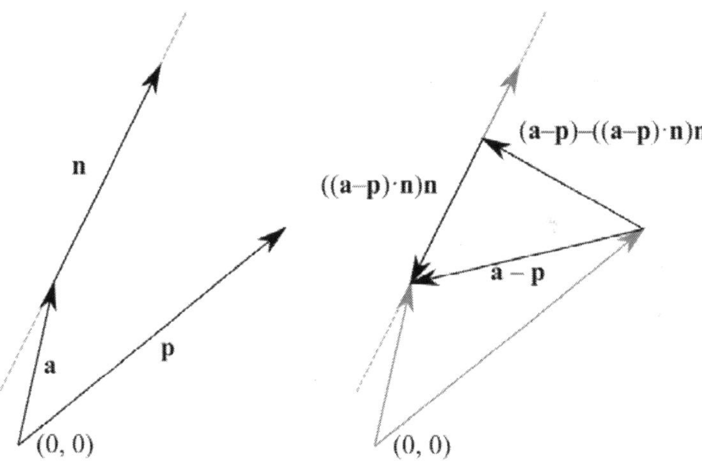

Figure: *Illustration of the vector formulation.*

Write the line in vector form:

$$\mathbf{x} = \mathbf{a} + t\mathbf{n}$$

where x is a 1×2 vector giving the two coordinate values of any arbitrary point on the line, n is a 1×2 unit vector in the direction of the line, a is a 1×2 vector giving the two coordinate dimensions of a particular point on the line, and t is a scalar. That is, a point x on the line is found by starting at a point a on the line, then moving t units along the direction of the line.

The distance of an arbitrary point p to this line is given by

$$\text{distance}(\mathbf{x} = \mathbf{a} + t\mathbf{n}, \mathbf{p}) = \| (\mathbf{a}-\mathbf{p}) - ((\mathbf{a}-\mathbf{p})\cdot\mathbf{n})\mathbf{n} \|.$$

This formula is constructed geometrically as follows: $\mathbf{a}-\mathbf{p}$ is a vector from p to the point a on the line. Then $(\mathbf{a}-\mathbf{p})\cdot\mathbf{n}$ is the projected length onto the line and so

$$((\mathbf{a}-\mathbf{p})\cdot\mathbf{n})\mathbf{n}$$

is a vector that is the projection of $\mathbf{a}-\mathbf{p}$ onto the line. Thus

$$(\mathbf{a}-\mathbf{p})-((\mathbf{a}-\mathbf{p})\cdot\mathbf{n})\mathbf{n}$$

is the component of $\mathbf{a}-\mathbf{p}$ perpendicular to the line. The distance from the point to the line is then just the norm of that vector. This more general formula can be used in dimensions other than two.

Another Vector Projection Proof

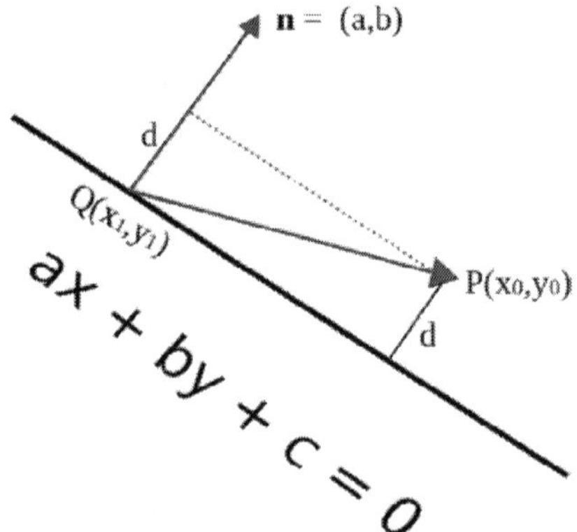

Let P be the point with coordinates (x_0, y_0) and let the given line have equation $ax + by + c = 0$. Also, let $Q = (x_1, y_1)$ be any point on this line and n the vector (a, b) starting at point Q. The vector n is perpendicular to the line, and the distance d from point P to the line is equal to the length of the orthogonal projection of $\overrightarrow{QP}$ on n. The length of this projection is given by:

$$d = \frac{|\overrightarrow{QP}\cdot\mathbf{n}|}{\|\mathbf{n}\|}.$$

Now,

$$\overrightarrow{QP} = (x_0 - x_1, y_0 - y_1), \text{ so}$$

$$\overrightarrow{QP}\cdot\mathbf{n} = a(x_0 - x_1) + b(y_0 - y_1) \text{ and } \|\mathbf{n}\| = \sqrt{a^2 + b^2},$$

thus

$$d = \frac{|a(x_0 - x_1) + b(y_0 - y_1)|}{\sqrt{a^2 + b^2}}.$$

Since Q is a point on the line, $c = -ax_1 - by_1$, and so,

$$d = \frac{|ax_0 + by_0 + c|}{\sqrt{a^2 + b^2}}.$$

Two Special Cases

In proofs that involve slopes of lines (as the ones below will), the main argument usually fails if the line in question is either vertical or horizontal. However, the result remains valid in these cases. In the general equation of a line, $ax + by + c = 0$, a and b can not both be zero. If $a = 0$, the line is horizontal and has equation $y = -c/b$. The distance from (x_0, y_0) to this line is measured along a vertical line segment of length $|y_0 - (-c/b)| = |by_0 + c| / |b|$ in accordance with the formula. Similarly, for vertical lines ($b = 0$) the distance between the same point and the line is $|ax_0 + c| / |a|$, as measured along a horizontal line segment.

An Algebraic Proof

This proof is only valid if the line is neither vertical nor horizontal, that is, we assume that neither a nor b in the equation of the line is zero.

The line with equation $ax + by + c = 0$ has slope $-a/b$, so any line perpendicular to it will have slope b/a (the negative reciprocal). Let (m, n) be the point of intersection of the line $ax + by + c = 0$ and the line perpendicular to it which passes through the point (x_0, y_0). The line through these two points is perpendicular to the original line, so

$$\frac{y_0 - n}{x_0 - m} = \frac{b}{a}.$$

Thus, $a(y_0 - n) - b(x_0 - m) = 0$, and by squaring this equation we obtain:

$$a^2(y_0 - n)^2 + b^2(x_0 - m)^2 = 2ab(y_0 - n)(x_0 - m).$$

Now consider,

$$(a(x_0 - m) + b(y_0 - n))^2 = a^2(x_0 - m)^2 + 2ab(y_0 - n)$$
$$(x_0 - m) + b^2(y_0 - n)^2 = (a^2 + b^2)((x_0 - m)^2 + (y_0 - n)^2)$$

using the above squared equation. But we also have,

$$(a(x_0 - m) + b(y_0 - n))^2 = (ax_0 + by_0 - am - bn)^2 = (ax_0 + by_0 + c)^2$$

since (m, n) is on $ax + by + c = 0$. Thus,

$$(a^2 + b^2)((x_0 - m)^2 + (y_0 - n)^2) = (ax_0 + by_0 + c)^2$$

and we obtain the length of the line segment determined by these two points,

$$d = \sqrt{(x_0 - m)^2 + (y_0 - n)^2} = \frac{|ax_0 + by_0 + c|}{\sqrt{a^2 + b^2}}.$$

A Geometric Proof

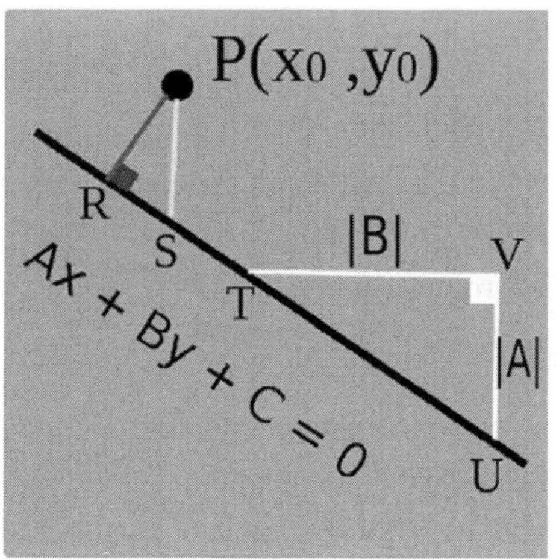

Figure: *Diagram for geometric proof*

This proof is valid only if the line is not horizontal or vertical.

Drop a perpendicular from the point *P* with coordinates (x_0, y_0) to the line with equation $Ax + By + C = 0$. Label the foot of the perpendicular *R*. Draw the vertical line through *P* and label its intersection with the given line *S*. At any point *T* on the line, draw a right triangle *TVU* whose sides are horizontal and vertical line segments with hypotenuse *TU* on the given line and horizontal side of length $|B|$. The vertical side of d*TVU* will have length $|A|$ since the line has slope $-A/B$.

d*SRP* and d*TVU* are similar triangles since they are both right triangles and d*PSR* Ed d*VUT* since they are corresponding angles of a transversal to the parallel lines *PS* and *UV* (both are vertical lines). Corresponding sides of these triangles are in the same ratio, so:

$$\frac{|\overline{PR}|}{|\overline{PS}|} = \frac{|\overline{TV}|}{|\overline{TU}|}.$$

If point *S* has coordinates (x_0, m) then $|PS| = |y_0 - m|$ and the distance from *P* to the line is:

$$|\overline{PR}| = \frac{|y_0 - m||B|}{\sqrt{A^2 + B^2}}.$$

Since S is on the line, we can find the value of m,

$$m = \frac{-Ax_0 - C}{B},$$

and finally obtain:[8]

$$|\overline{PR}| = \frac{|Ax_0 + By_0 + C|}{\sqrt{A^2 + B^2}}.$$

Another Formula

It is possible to produce another expression to find the shortest distance of a point to a line. This derivation also requires that the line is not vertical or horizontal.

The point P is given with coordinates (x_0, y_0). The equation of a line is given by $y = mx + k$. The equation of the normal of that line which passes through the point P is given $y = \frac{x_0 - x}{m} + y_0$.

The point at which these two lines intersect is the closest point on the original line to the point P. Hence:

$$mx + k = \frac{x_0 - x}{m} + y_0.$$

We can solve this equation for x,

$$x = \frac{x_0 + my_0 - mk}{m^2 + 1}.$$

The y coordinate of the point of intersection can be found by substituting this value of x into the equation of the original line,

$$y = m\frac{(x_0 + my_0 - mk)}{m^2 + 1} + k.$$

Using the equation for finding the distance between 2 points, $d = \sqrt{(X_2 - X_1)^2 + (Y_2 - Y_1)^2}$, we can deduce that the formula to find the shortest distance between a line and a point is the following:

$$d = \sqrt{\left(\frac{x_0 + my_0 - mk}{m^2 + 1} - x_0\right)^2 + \left(m\frac{x_0 + my_0 - mk}{m^2 + 1} + k - y_0\right)^2}.$$

Recalling that $m = -a/b$ and $k = -c/b$ for the line with equation $ax + by + c = 0$, a little algebraic simplification reduces this to the standard expression.

Line Defined by Two Points

If the line passes through two points, (x_1, y_1) and (x_2, y_2), and if we write D_x for $(x_2 - x_1)$ and D_y for $(y_2 - y_1)$, the perpendicular distance from (x_0, y_0) to the line is given by:

$$d = \frac{|D_y x_0 - D_x y_0 - x_1 y_2 + x_2 y_1|}{\sqrt{(D_x)^2 + (D_y)^2}}$$

The denominator of this expression is the distance between (x_1,y_1) and (x_2,y_2). The numerator is twice the area of the triangle with its vertices at the three points, (x_0,y_0), (x_1,y_1), and (x_2,y_2). The expression is equivalent to $h = \frac{2A}{b}$, which can be obtained by rearranging the standard formula for the area of a triangle: $A = \frac{1}{2}bh$, where b is the length of a side, and h is the perpendicular height from the opposite vertex.

Triangle

A triangle is a polygon with three edges and three vertices. It is one of the basic shapes in geometry. A triangle with vertices *A*, *B*, and *C* is denoted $\triangle ABC$.

In Euclidean geometry any three points, when non-collinear, determine a unique triangle and a unique plane (i.e. a two-dimensional Euclidean space).

Types of Triangle

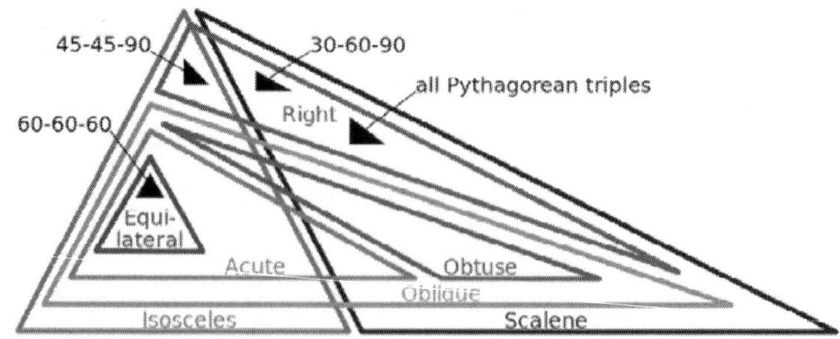

Figure: *Euler diagram of types of triangles, using the definition that isosceles triangles have* at least *2 equal sides, i.e. equilateral triangles are isosceles.*

By Relative Lengths of Sides

Triangles can be classified according to the relative lengths of their sides:

- In an equilateral triangle all sides have the same length. An equilateral triangle is also a regular polygon with all angles measuring 60°.
- In an isosceles triangle, two sides are equal in length. An isosceles triangle also has two angles of the same measure; namely, the angles opposite to the two sides of the same length; this fact is the content of the isosceles triangle theorem, which was known by Euclid. Some mathematicians define an isosceles triangle to have exactly two equal sides, whereas others define an isosceles triangle as one with *at least* two equal sides. The latter definition would make all equilateral triangles isosceles triangles. The 45–45–90 right triangle, which appears in the tetrakis square tiling, is isosceles.
- In a scalene triangle, all sides are unequal, and equivalently all angles are unequal. A right triangle is also a scalene triangle if and only if it is not isosceles.

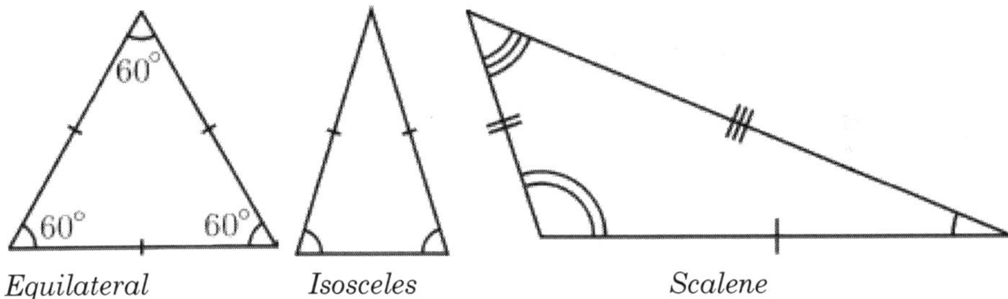

Equilateral *Isosceles* *Scalene*

Hatch marks, also called tick marks, are used in diagrams of triangles and other geometric figures to identify sides of equal lengths. A side can be marked with a pattern of "ticks", short line segments in the form of tally marks; two sides have equal lengths if they are both marked with the same pattern. In a triangle, the pattern is usually no more than 3 ticks. An equilateral triangle has the same pattern on all 3 sides, an isosceles triangle has the same pattern on just 2 sides, and a scalene triangle has different patterns on all sides since no sides are equal. Similarly, patterns of 1, 2, or 3 concentric arcs inside the angles are used to indicate equal angles. An equilateral triangle has the same pattern on all 3 angles, an isosceles triangle has the same pattern on just 2 angles, and a scalene triangle has different patterns on all angles since no angles are equal.

By Internal Angles

Triangles can also be classified according to their internal angles, measured here in degrees.

- A right triangle (or *right-angled triangle,* formerly called a *rectangled triangle*) has one of its interior angles measuring 90° (a right angle). The side opposite to the right angle is the hypotenuse, the longest side of the triangle. The other two sides are called the *legs* or *catheti* (singular: cathetus) of the triangle. Right triangles obey the Pythagorean theorem: the sum of the squares of the lengths of the two legs is equal to the square of the length of the hypotenuse: $a^2 + b^2 = c^2$, where a and b are the lengths of the legs and c is the length of the hypotenuse. Special right triangles are right triangles with additional properties that make calculations involving them easier. One of the two most famous is the 3–4–5 right triangle, where $3^2 + 4^2 = 5^2$. In this situation, 3, 4, and 5 are a Pythagorean triple. The other one is an isosceles triangle that has 2 angles that each measure 45 degrees.
- Triangles that do not have an angle measuring 90° are called oblique triangles.
- A triangle with all interior angles measuring less than 90° is an acute triangle or *acute-angled triangle.* If c is the length of the longest side, then $a^2 + b^2 > c^2$, where a and b are the lengths of the other sides.
- A triangle with one interior angle measuring more than 90° is an obtuse triangle or *obtuse-angled triangle.* If c is the length of the longest side, then $a^2 + b^2 < c^2$, where a and b are the lengths of the other sides.
- A triangle with an interior angle of 180° (and collinear vertices) is degenerate.
- A right degenerate triangle has collinear vertices, two of which are coincident.

A triangle that has two angles with the same measure also has two sides with the same length, and therefore it is an isosceles triangle. It follows that in a triangle where all angles have the same measure, all three sides have the same length, and such a triangle is therefore equilateral.

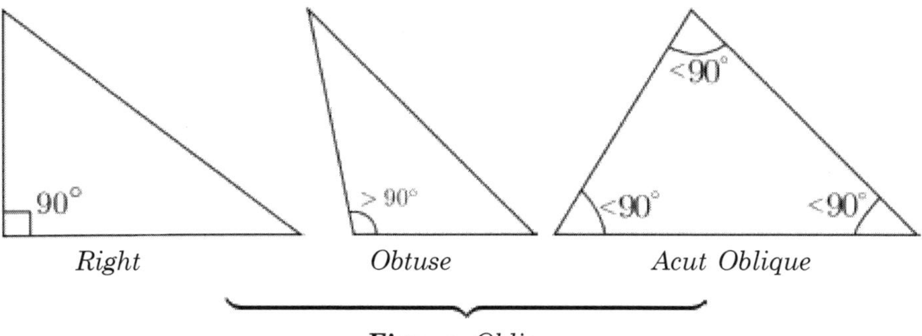

Figure: *Oblique*

Basic Facts

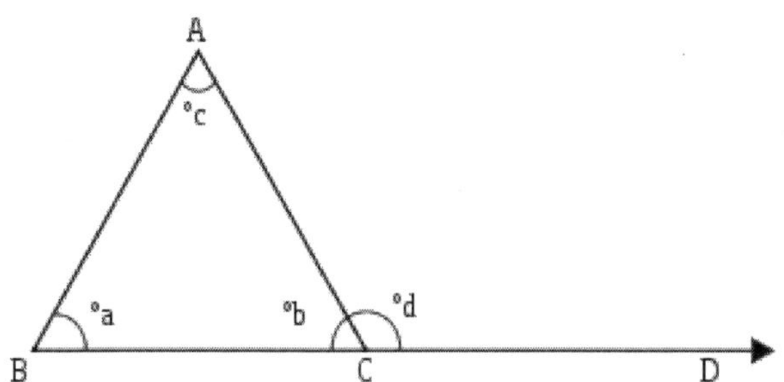

Figure: *A triangle, showing exterior angle d.*

Triangles are assumed to be two-dimensional plane figures, unless the context provides otherwise. In rigorous treatments, a triangle is therefore called a *2*-simplex. Elementary facts about triangles were presented by Euclid in books 1–4 of his Elements, around 300 BC.

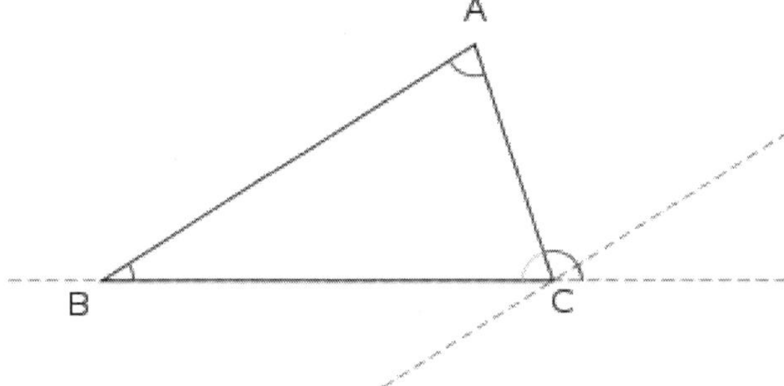

Figure: *The measures of the interior angles of the triangle always add up to 180 degrees (same colour to point out they are equal).*

The measures of the interior angles of a triangle in Euclidean space always add up to 180 degrees. This allows determination of the measure of the third angle of any triangle given the measure of two angles. An exterior angle of a triangle is an angle that is a linear pair (and hence supplementary) to an interior angle. The measure of an exterior angle of a triangle is equal to the sum of the measures of the two interior angles that are not adjacent to it; this is the exterior angle theorem. The sum of the measures of the three exterior angles (one for each vertex) of any triangle is 360 degrees.

Similarity and Congruence

Two triangles are said to be similar if every angle of one triangle has the same measure as the corresponding angle in the other triangle. The corresponding sides of similar triangles have lengths that are in the same proportion, and this property is also sufficient to establish similarity.

Some basic theorems about similar triangles are:

- If two corresponding internal angles of two triangles have the same measure, the triangles are similar.
- If two corresponding sides of two triangles are in proportion, and their included angles have the same measure, then the triangles are similar. (The *included angle* for any two sides of a polygon is the internal angle between those two sides.)
- If three corresponding sides of two triangles are in proportion, then the triangles are similar.

Two triangles that are congruent have exactly the same size and shape:[note 4] all pairs of corresponding interior angles are equal in measure, and all pairs of corresponding sides have the same length. (This is a total of six equalities, but three are often sufficient to prove congruence.)

Some sufficient conditions for a pair of triangles to be congruent are:

- SAS Postulate: Two sides in a triangle have the same length as two sides in the other triangle, and the included angles have the same measure.
- ASA: Two interior angles and the included side in a triangle have the same measure and length, respectively, as those in the other triangle. (The *included side* for a pair of angles is the side that is common to them.)
- SSS: Each side of a triangle has the same length as a corresponding side of the other triangle.
- AAS: Two angles and a corresponding (non-included) side in a triangle have the same measure and length, respectively, as

those in the other triangle. (This is sometimes referred to as *AAcorrS* and then includes ASA above.)

- Hypotenuse-Leg (HL) Theorem: The hypotenuse and a leg in a right triangle have the same length as those in another right triangle. This is also called RHS (right-angle, hypotenuse, side).
- Hypotenuse-Angle Theorem: The hypotenuse and an acute angle in one right triangle have the same length and measure, respectively, as those in the other right triangle. This is just a particular case of the AAS theorem.

An important condition is:

- Side-Side-Angle (or Angle-Side-Side) condition: If two sides and a corresponding non-included angle of a triangle have the same length and measure, respectively, as those in another triangle, then this is *not* sufficient to prove congruence; but if the angle given is opposite to the longer side of the two sides, then the triangles are congruent. The Hypotenuse-Leg Theorem is a particular case of this criterion. The Side-Side-Angle condition does not by itself guarantee that the triangles are congruent because one triangle could be obtuse-angled and the other acute-angled.

Using right triangles and the concept of similarity, the trigonometric functions sine and cosine can be defined. These are functions of an angle which are investigated in trigonometry.

Right Triangles

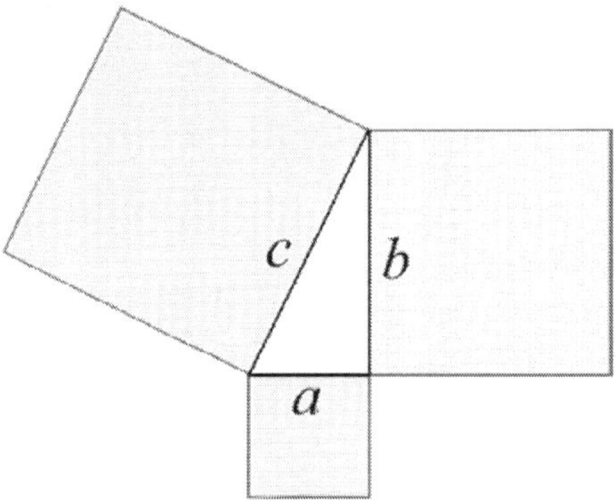

Figure: *The Pythagorean theorem*

A central theorem is the Pythagorean theorem, which states in any right triangle, the square of the length of the hypotenuse equals the sum of the squares of the lengths of the two other sides. If the hypotenuse has length c, and the legs have lengths a and b, then the theorem states that

$$a^2 + b^2 = c^2.$$

The converse is true: if the lengths of the sides of a triangle satisfy the above equation, then the triangle has a right angle opposite side c.

Some other facts about right triangles:

- The acute angles of a right triangle are complementary.

$$a + b + 90^\circ = 180^\circ \Rightarrow a + b = 90^\circ \Rightarrow a = 90^\circ - b$$

- If the legs of a right triangle have the same length, then the angles opposite those legs have the same measure. Since these angles are complementary, it follows that each measures 45 degrees. By the Pythagorean theorem, the length of the hypotenuse is the length of a leg times √2.
- In a right triangle with acute angles measuring 30 and 60 degrees, the hypotenuse is twice the length of the shorter side, and the longer side is equal to the length of the shorter side times √3:

$$c = 2a$$

$$b = a \times \sqrt{3}.$$

For all triangles, angles and sides are related by the law of cosines and law of sines (also called the *cosine rule* and *sine rule*).

Existence of a Triangle

The triangle inequality states that the sum of the lengths of any two sides of a triangle must be greater than or equal to the length of the third side. That sum can equal the length of the third side only in the case of a degenerate triangle, one with collinear vertices. It is not possible for that sum to be less than the length of the third side.

Trigonometric Conditions

Three positive angles α, β, and γ, each of them less than 180°, are the angles of a triangle if and only if any one of the following conditions holds:

$$\tan\frac{\alpha}{2}\tan\frac{\beta}{2} + \tan\frac{\beta}{2}\tan\frac{\gamma}{2} + \tan\frac{\gamma}{2}\tan\frac{\alpha}{2} = 1,$$

$$\sin^2\frac{\alpha}{2} + \sin^2\frac{\beta}{2} + \sin^2\frac{\gamma}{2} + 2\sin\frac{\alpha}{2}\sin\frac{\beta}{2}\sin\frac{\gamma}{2} = 1.$$

Points, Lines, and Circles Associated with a Triangle

There are hundreds of different constructions that find a special point associated with (and often inside) a triangle, satisfying some unique property: Often they are constructed by finding three lines associated in a symmetrical way with the three sides (or vertices) and then proving that the three lines meet in a single point: an important tool for proving the existence of these is Ceva's theorem, which gives a criterion for determining when three such lines are concurrent. Similarly, lines associated with a triangle are often constructed by proving that three symmetrically constructed points are collinear: here Menelaus' theorem gives a useful general criterion. In this section just a few of the most commonly encountered constructions are explained.

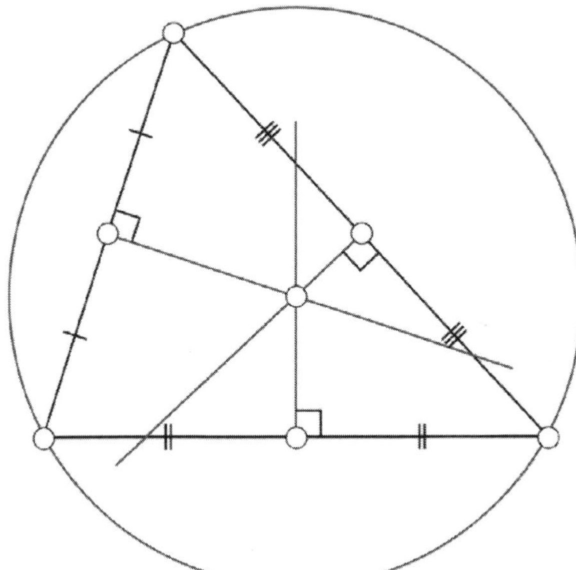

Figure: *The circumcenter is the center of a circle passing through the three vertices of the triangle.*

A perpendicular bisector of a side of a triangle is a straight line passing through the midpoint of the side and being perpendicular to it, i.e. forming a right angle with it. The three perpendicular bisectors meet in a single point, the triangle's circumcenter, usually denoted by O; this point is the center of the circumcircle, the circle passing through all three vertices. The diameter of this circle, called the *circumdiameter*, can be found from the law of sines stated above. The circumcircle's radius is called the *circumradius*.

Thales' theorem implies that if the circumcenter is located on one side of the triangle, then the opposite angle is a right one. If the circumcenter is located inside the triangle, then the triangle is acute; if the circumcenter is located outside the triangle, then the triangle is obtuse.

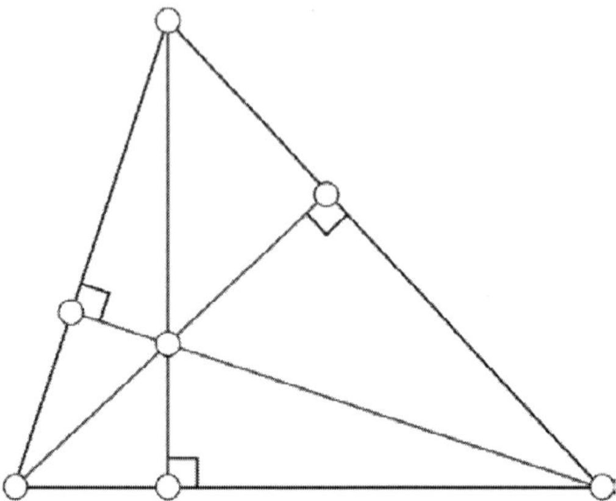

Figure: *The intersection of the altitudes is the orthocenter.*

An altitude of a triangle is a straight line through a vertex and perpendicular to (i.e. forming a right angle with) the opposite side. This opposite side is called the *base* of the altitude, and the point where the altitude intersects the base (or its extension) is called the *foot* of the altitude. The length of the altitude is the distance between the base and the vertex. The three altitudes intersect in a single point, called the orthocenter of the triangle, usually denoted by H. The orthocenter lies inside the triangle if and only if the triangle is acute.

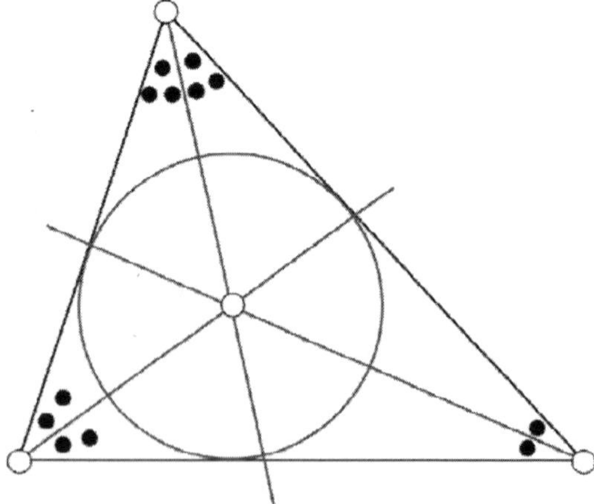

Figure: *The intersection of the angle bisectors is the center of the incircle.*

An angle bisector of a triangle is a straight line through a vertex which cuts the corresponding angle in half. The three angle bisectors intersect in a single point, the incenter, usually denoted by I, the center of the triangle's incircle. The incircle is the circle which lies inside the triangle and touches all three sides. Its radius is called the *inradius*. There are three other important circles, the excircles; they lie outside the triangle and touch one side as well as the extensions of the other two. The centers of the in- and excircles form an orthocentric system.

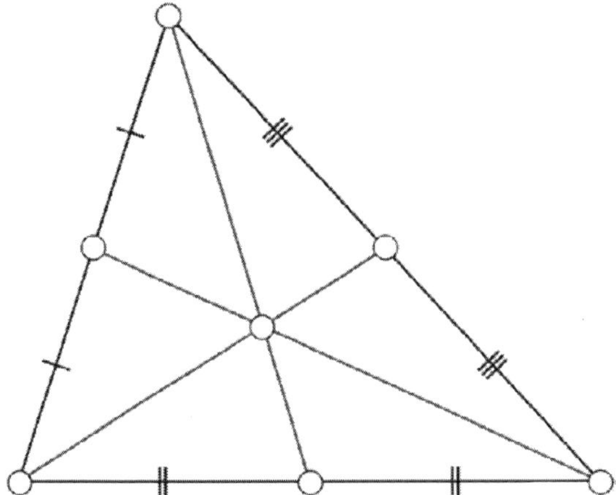

Figure; *The intersection of the medians is the centroid.*

A median of a triangle is a straight line through a vertex and the midpoint of the opposite side, and divides the triangle into two equal areas. The three medians intersect in a single point, the triangle's centroid or geometric barycenter, usually denoted by G. The centroid of a rigid triangular object (cut out of a thin sheet of uniform density) is also its center of mass: the object can be balanced on its centroid in a uniform gravitational field. The centroid cuts every median in the ratio 2:1, i.e. the distance between a vertex and the centroid is twice the distance between the centroid and the midpoint of the opposite side.

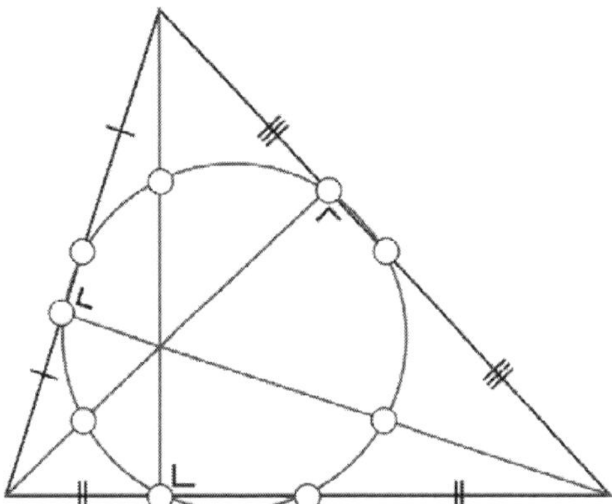

Figure: *Nine-point circle demonstrates a symmetry where six points lie on the edge of the triangle.*

The midpoints of the three sides and the feet of the three altitudes all lie on a single circle, the triangle's nine-point circle. The remaining three points for which it is named are the midpoints of the portion of altitude between the vertices and the orthocenter. The radius of the nine-point circle is half that of the circumcircle. It touches the incircle (at the Feuerbach point) and the three excircles.

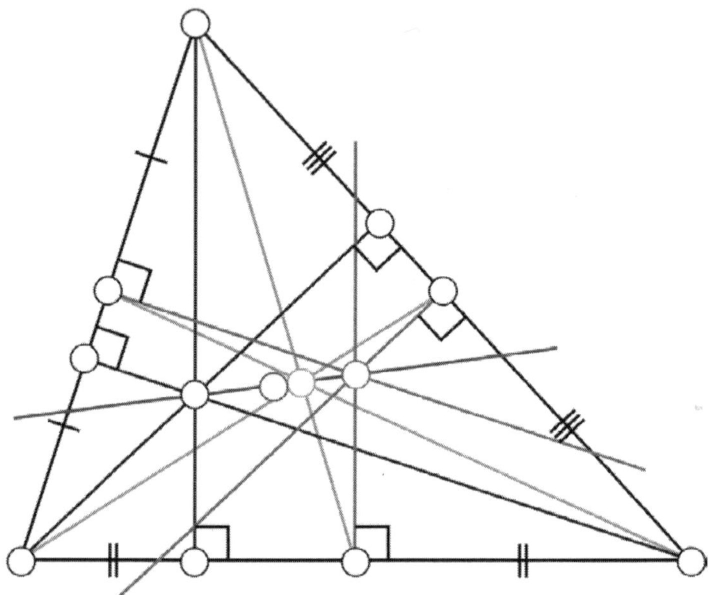

Figure: *Euler's line is a straight line through the centroid, orthocenter, circumcenter (green) and center of the nine-point circle.*

The centroid, orthocenter, circumcenter and center of the nine-point circle (red point) all lie on a single line, known as Euler's line. The center of the nine-point circle lies at the midpoint between the orthocenter and the circumcenter, and the distance between the centroid and the circumcenter is half that between the centroid and the orthocenter.

The center of the incircle is not in general located on Euler's line.

If one reflects a median in the angle bisector that passes through the same vertex, one obtains a symmedian. The three symmedians intersect in a single point, the symmedian point of the triangle.

Computing the Sides and Angles

There are various standard methods for calculating the length of a side or the measure of an angle. Certain methods are suited to calculating values in a right-angled triangle; more complex methods may be required in other situations.

Trigonometric Ratios in Right Triangles

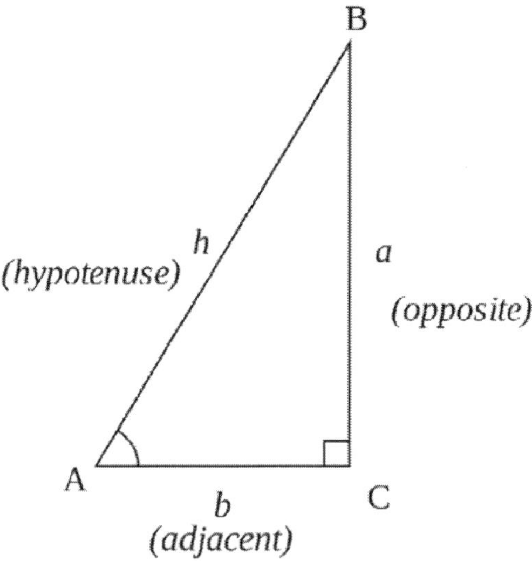

A right triangle always includes a 90° (π/2 radians) angle, here with label C. Angles A and B may vary. Trigonometric functions specify the relationships among side lengths and interior angles of a right triangle.

In right triangles, the trigonometric ratios of sine, cosine and tangent can be used to find unknown angles and the lengths of unknown sides. The sides of the triangle are known as follows:

- The hypotenuse is the side opposite the right angle, or defined as the longest side of a right-angled triangle, in this case h.
- The *opposite side* is the side opposite to the angle we are interested in, in this case a.
- The *adjacent side* is the side that is in contact with the angle we are interested in and the right angle, hence its name. In this case the adjacent side is b.

Sine, Cosine and Tangent

The *sine* of an angle is the ratio of the length of the opposite side to the length of the hypotenuse. In our case

$$\sin A = \frac{\text{oppositeside}}{\text{hypotenuse}} = \frac{a}{h}.$$

Note that this ratio does not depend on the particular right triangle chosen, as long as it contains the angle A, since all those triangles are similar.

The *cosine* of an angle is the ratio of the length of the adjacent side to the length of the hypotenuse. In our case

$$\cos A = \frac{\text{adjacentside}}{\text{hypotenuse}} = \frac{b}{h}.$$

The *tangent* of an angle is the ratio of the length of the opposite side to the length of the adjacent side. In our case

$$\tan A = \frac{\text{oppositeside}}{\text{adjacentside}} = \frac{a}{b} = \frac{\sin A}{\cos A}.$$

The acronym "SOH-CAH-TOA" is a useful mnemonic for these ratios.

Inverse Functions

The inverse trigonometric functions can be used to calculate the internal angles for a right angled triangle with the length of any two sides.

Arcsin can be used to calculate an angle from the length of the opposite side and the length of the hypotenuse.

$$\theta = \arcsin\left(\frac{\text{opposite side}}{\text{hypotenuse}}\right)$$

Arccos can be used to calculate an angle from the length of the adjacent side and the length of the hypontenuse.

$$\theta = \arccos\left(\frac{\text{adjacent side}}{\text{hypotenuse}}\right)$$

Arctan can be used to calculate an angle from the length of the opposite side and the length of the adjacent side.

$$\theta = \arctan\left(\frac{\text{opposite side}}{\text{adjacent side}}\right)$$

In introductory geometry and trigonometry courses, the notation $\sin^{-1}$, $\cos^{-1}$, etc., are often used in place of arcsin, arccos, etc. However, the arcsin, arccos, etc., notation is standard in higher mathematics where trigonometric functions are commonly raised to powers, as this avoids confusion between multiplicative inverse and compositional inverse.

Sine, Cosine and Tangent Rules

The law of sines, or sine rule, states that the ratio of the length of a side to the sine of its corresponding opposite angle is constant, that is

$$\frac{a}{\sin\alpha}=\frac{b}{\sin\beta}=\frac{c}{\sin\gamma}.$$

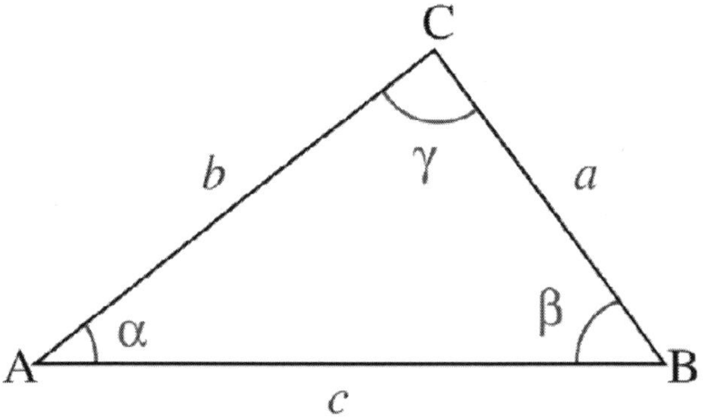

Figure: *A triangle with sides of length a, b and c and angles of α, β and γ respectively.*

This ratio is equal to the diameter of the circumscribed circle of the given triangle. Another interpretation of this theorem is that every triangle with angles α, β and γ is similar to a triangle with side lengths equal to sin α, sin β and sin γ. This triangle can be constructed by first constructing a circle of diameter 1, and inscribing in it two of the angles of the triangle. The length of the sides of that triangle will be sin α, sin β and sin γ. The side whose length is sin α is opposite to the angle whose measure is α, etc.

The law of cosines, or cosine rule, connects the length of an unknown side of a triangle to the length of the other sides and the angle opposite to the unknown side. As per the law:

For a triangle with length of sides a, b, c and angles of α, β, γ respectively, given two known lengths of a triangle a and b, and the angle between the two known sides γ (or the angle opposite to the unknown side c), to calculate the third side c, the following formula can be used:

$$c^2=a^2+b^2-2ab\cos(\gamma)$$

$$b^2=a^2+c^2-2ac\cos(\beta)$$

$$a^2=b^2+c^2-2bc\cos(\alpha)$$

If the lengths of all three sides of any triangle are known the three angles can be calculated:

$$\alpha=\arccos\left(\frac{b^2+c^2-a^2}{2bc}\right)$$

$$\beta = \arccos\left(\frac{a^2 + c^2 - b^2}{2ac}\right)$$

$$\gamma = \arccos\left(\frac{a^2 + b^2 - c^2}{2ab}\right)$$

The law of tangents or tangent rule, can be used to find a side or an angle when you know two sides and an angle or two angles and a side. It states that:

$$\frac{a-b}{a+b} = \frac{\tan[\frac{1}{2}(\alpha - \beta)]}{\tan[\frac{1}{2}(\alpha + \beta)]}.$$

Solution of Triangles

"Solution of triangles" is the historical term for the solving of the main trigonometric problem: to find missing characteristics of a triangle (three angles, the lengths of the three sides etc.) when at least three of these characteristics are given. The triangle can be located on a plane or on a sphere. This problem often occurs in various trigonometric applications, such as geodesy, astronomy, construction, navigation etc.

Computing the Area of a Triangle

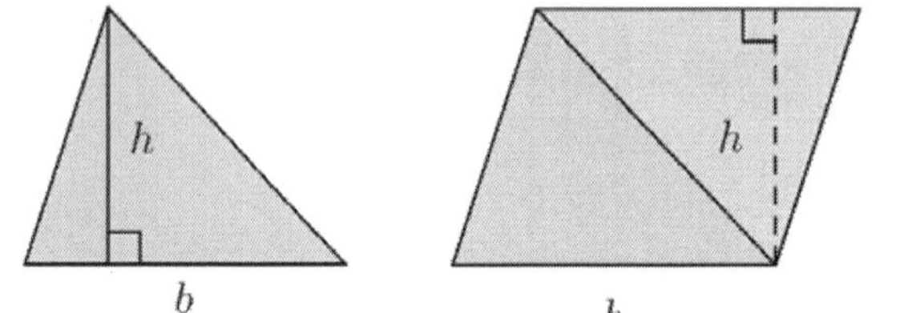

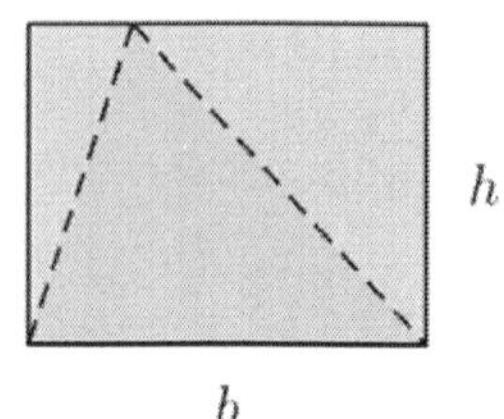

Figure: *The area of a triangle can be demonstrated as half of the area of a parallelogram which has the same base length and height.*

Calculating the area T of a triangle is an elementary problem encountered often in many different situations. The best known and simplest formula is:

$$T = \frac{1}{2}bh$$

where b is the length of the base of the triangle, and h is the height or altitude of the triangle. The term "base" denotes any side, and "height" denotes the length of a perpendicular from the vertex opposite the side onto the line containing the side itself. In 499 CE Aryabhata, a great mathematician-astronomer from the classical age of Indian mathematics and Indian astronomy, used this method in the Aryabhatiya.

Although simple, this formula is only useful if the height can be readily found, which is not always the case. For example, the surveyor of a triangular field might find it relatively easy to measure the length of each side, but relatively difficult to construct a 'height'. Various methods may be used in practice, depending on what is known about the triangle. The following is a selection of frequently used formulae for the area of a triangle.

Using Trigonometry

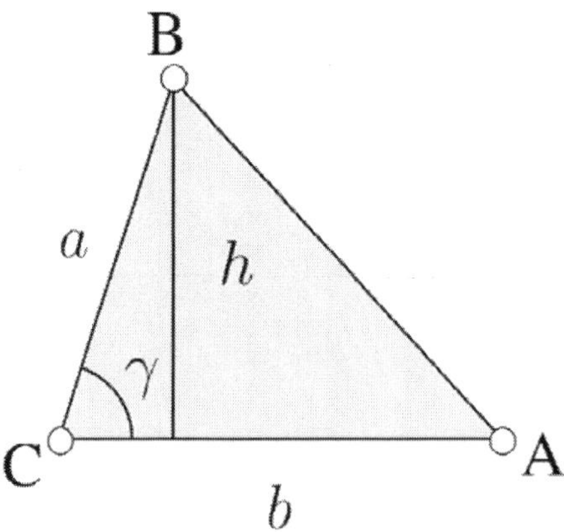

Figure: *Applying trigonometry to find the altitude* h.

The height of a triangle can be found through the application of trigonometry.

Knowing SAS: Using the labels in the image on the right, the altitude is $h = a \sin \gamma$. Substituting this in the formula $T = \frac{1}{2}bh$ derived above, the area of the triangle can be expressed as:

$$T = \frac{1}{2}ab\sin\gamma = \frac{1}{2}bc\sin\alpha = \frac{1}{2}ca\sin\beta$$

(where α is the interior angle at *A*, β is the interior angle at *B*, is the interior angle at *C* and *c* is the line AB).

Furthermore, since sin α = sin (π – α) = sin (β + γ), and similarly for the other two angles:

$$T = \frac{1}{2}ab\sin(\alpha+\beta) = \frac{1}{2}bc\sin(\beta+\gamma) = \frac{1}{2}ca\sin(\gamma+\alpha).$$

Knowing AAS:

$$T = \frac{b^2(\sin\alpha)(\sin(\alpha+\beta))}{2\sin\beta},$$

and analogously if the known side is a or c.

Knowing ASA:

$$T = \frac{a^2}{2(\cot\beta+\cot\gamma)} = \frac{a^2(\sin\beta)(\sin\gamma)}{2\sin(\beta+\gamma)},$$

and analogously if the known side is b or c.

Using Heron's Formula

The shape of the triangle is determined by the lengths of the sides. Therefore the area can also be derived from the lengths of the sides. By Heron's formula:

$$T = \sqrt{s(s-a)(s-b)(s-c)}$$

where $s = \frac{a+b+c}{2}$ is the *semiperimeter*, or half of the triangle's perimeter.

Three other equivalent ways of writing Heron's formula are

$$T = \frac{1}{4}\sqrt{(a^2+b^2+c^2)^2 - 2(a^4+b^4+c^4)}$$

$$T = \frac{1}{4}\sqrt{(a+b-c)(a-b+c)(-a+b+c)(a+b+c)}.$$

Using Vectors

The area of a parallelogram embedded in a three-dimensional Euclidean space can be calculated using vectors. Let vectors AB and AC point respectively from A to B and from A to C. The area of parallelogram $ABDC$ is then

$$|\mathbf{AB}\times\mathbf{AC}|,$$

which is the magnitude of the cross product of vectors AB and AC. The area of triangle ABC is half of this,

$$\frac{1}{2}|\mathbf{AB}\times\mathbf{AC}|.$$

The area of triangle ABC can also be expressed in terms of dot products as follows:

$$\frac{1}{2}\sqrt{(\mathbf{AB}\cdot\mathbf{AB})(\mathbf{AC}\cdot\mathbf{AC})-(\mathbf{AB}\cdot\mathbf{AC})^2} = \frac{1}{2}\sqrt{|\mathbf{AB}|^2|\mathbf{AC}|^2-(\mathbf{AB}\cdot\mathbf{AC})^2}.$$

In two-dimensional Euclidean space, expressing vector AB as a free vector in Cartesian space equal to (x_1,y_1) and AC as (x_2,y_2), this can be rewritten as:

$$\frac{1}{2}|x_1y_2 - x_2y_1|.$$

Using Coordinates

If vertex A is located at the origin (0, 0) of a Cartesian coordinate system and the coordinates of the other two vertices are given by $B = (x_B, y_B)$ and $C = (x_C, y_C)$, then the area can be computed as $^1/_2$ times the absolute value of the determinant

$$T = \frac{1}{2}\left|\det\begin{pmatrix} x_B & x_C \\ y_B & y_C \end{pmatrix}\right| = \frac{1}{2}|x_By_C - x_Cy_B|.$$

For three general vertices, the equation is:

$$T = \frac{1}{2}\left|\det\begin{pmatrix} x_A & x_B & x_C \\ y_A & y_B & y_C \\ 1 & 1 & 1 \end{pmatrix}\right| = \frac{1}{2}\left|x_Ay_B - x_Ay_C + x_By_C - x_By_A + x_Cy_A - x_Cy_B\right|,$$

which can be written as

$$T = \frac{1}{2}\left|(x_A - x_C)(y_B - y_A) - (x_A - x_B)(y_C - y_A)\right|.$$

If the points are labelled sequentially in the counterclockwise direction, the above determinant expressions are positive and the absolute value signs can be omitted. The above formula is known as the shoelace formula or the surveyor's formula.

If we locate the vertices in the complex plane and denote them in counterclockwise sequence as $a = x_A + y_Ai$, $b = x_B + y_Bi$, and $c = x_C + y_Ci$, and denote their complex conjugates as $\bar{a}$, $\bar{b}$, and $\bar{c}$, then the formula

$$T = \frac{i}{4}\begin{vmatrix} a & \bar{a} & 1 \\ b & \bar{b} & 1 \\ c & \bar{c} & 1 \end{vmatrix}$$

is equivalent to the shoelace formula.

In three dimensions, the area of a general triangle $A = (x_A, y_A, z_A)$, $B = (x_B, y_B, z_B)$ and $C = (x_C, y_C, z_C)$ is the Pythagorean sum of the areas of the respective projections on the three principal planes (i.e. $x = 0$, $y = 0$ and $z = 0$):

$$T=\frac{1}{2}\sqrt{\begin{vmatrix} x_A & x_B & x_C \\ y_A & y_B & y_C \\ 1 & 1 & 1 \end{vmatrix}^2 + \begin{vmatrix} y_A & y_B & y_C \\ z_A & z_B & z_C \\ 1 & 1 & 1 \end{vmatrix}^2 + \begin{vmatrix} z_A & z_B & z_C \\ x_A & x_B & x_C \\ 1 & 1 & 1 \end{vmatrix}^2}.$$

Using Line Integrals

The area within any closed curve, such as a triangle, is given by the line integral around the curve of the algebraic or signed distance of a point on the curve from an arbitrary oriented straight line L. Points to the right of L as oriented are taken to be at negative distance from L, while the weight for the integral is taken to be the component of arc length parallel to L rather than arc length itself.

This method is well suited to computation of the area of an arbitrary polygon. Taking L to be the x-axis, the line integral between consecutive vertices (x_i, y_i) and (x_{i+1}, y_{i+1}) is given by the base times the mean height, namely $(x_{i+1}$ d $x_i)(y_i + y_{i+1})/2$. The sign of the area is an overall indicator of the direction of traversal, with negative area indicating counterclockwise traversal. The area of a triangle then falls out as the case of a polygon with three sides.

While the line integral method has in common with other coordinate-based methods the arbitrary choice of a coordinate system, unlike the others it makes no arbitrary choice of vertex of the triangle as origin or of side as base. Furthermore the choice of coordinate system defined by L commits to only two degrees of freedom rather than the usual three, since the weight is a local distance (e.g. x_{i+1} d x_i in the above) whence the method does not require choosing an axis normal to L.

When working in polar coordinates it is not necessary to convert to cartesian coordinates to use line integration, since the line integral between consecutive vertices (r_i, θ_i) and (r_{i+1}, θ_{i+1}) of a polygon is given directly by $r_i r_{i+1} \sin(\theta_{i+1}$ d $\theta_i)/2$. This is valid for all values of θ, with some decrease in numerical accuracy when $|\theta|$ is many orders of magnitude greater than π. With this formulation negative area indicates clockwise traversal, which should be kept in mind when mixing polar and cartesian coordinates. Just as the choice of y-axis $(x = 0)$ is immaterial for line integration in cartesian coordinates, so is the choice of zero heading $(\theta = 0)$ immaterial here.

Formulas Resembling Heron's Formula

Three formulas have the same structure as Heron's formula but are expressed in terms of different variables. First, denoting the medians from sides a, b, and c respectively as m_a, m_b, and m_c and their semi-sum $(m_a + m_b + m_c)/2$ as σ, we have

$$T = \frac{4}{3}\sqrt{\sigma(\sigma - m_a)(\sigma - m_b)(\sigma - m_c)}.$$

Next, denoting the altitudes from sides a, b, and c respectively as h_a, h_b, and h_c, and denoting the semi-sum of the reciprocals of the altitudes as $H = (h_a^{-1} + h_b^{-1} + h_c^{-1})/2$ we have

$$T^{-1} = 4\sqrt{H(H - h_a^{-1})(H - h_b^{-1})(H - h_c^{-1})}.$$

And denoting the semi-sum of the angles' sines as $S = [(\sin\alpha) + (\sin\beta) + (\sin\gamma)]/2$, we have

$$T = D^2\sqrt{S(S - \sin\alpha)(S - \sin\beta)(S - \sin\gamma)}$$

where D is the diameter of the circumcircle: $D = \frac{a}{\sin\alpha} = \frac{b}{\sin\beta} = \frac{c}{\sin\gamma}$.

Using Pick's Theorem

The theorem states:

$$T = I + \frac{1}{2}B - 1$$

where I is the number of internal lattice points and B is the number of lattice points lying on the border of the polygon.

Other Area Formulas

Numerous other area formulas exist, such as

$$T = r \cdot s,$$

where r is the inradius, and s is the semiperimeter (in fact this formula holds for *all* tangential polygons);

$$T = \frac{1}{2}D^2(\sin\alpha)(\sin\beta)(\sin\gamma)$$

and

$$T = \frac{abc}{2D} = \frac{abc}{4R}$$

for circumdiameter D; and[16]

$$T = \frac{\tan\alpha}{4}(b^2 + c^2 - a^2)$$

for angle $\alpha \neq 90°$.

Denoting the radius of the inscribed circle as r and the radii of the excircles as r_1, r_2, and r_3, the area can be expressed as

$$T = \sqrt{r r_1 r_2 r_3}.$$

In 1885, Baker[18] gave a collection of over a hundred distinct area formulas for the triangle. These include:

$$T = \frac{1}{2}[abch_a h_b h_c]^{1/3},$$

$$T = \frac{1}{2}\sqrt{abh_a h_b},$$

$$T = \frac{a+b}{2(h_a^{-1} + h_b^{-1})},$$

$$T = \frac{Rh_b h_c}{a}$$

for circumradius (radius of the circumcircle) R, and

$$T = \frac{h_a h_b}{2\sin\gamma}.$$

Upper Bound on the Area

The area of any triangle with perimeter p is less than or equal to $\frac{p^2}{12\sqrt{3}}$, with equality holding if and only if the triangle is equilateral.

Other upper bounds on the area T are given by

$$4\sqrt{3}T \le a^2 + b^2 + c^2$$

and

$$4\sqrt{3}T \le \frac{9abc}{a+b+c},$$

both again holding if and only if the triangle is equilateral.

Bisecting the Area

There are infinitely many lines that bisect the area of a triangle. Three of them are the medians, which are the only area bisectors that go through the centroid. Three other area bisectors are parallel to the triangle's sides.

Any line through a triangle that splits both the triangle's area and its perimeter in half goes through the triangle's incenter. There can be one, two, or three of these for any given triangle.

Line

The notion of line or straight line was introduced by ancient mathematicians to represent straight objects with negligible width and depth. Lines are an idealization of such objects. Until the seventeenth century, lines were defined like this: "The line is the first species of quantity, which has only one dimension, namely length, without any width nor depth, and is nothing else than the flow or run of the point which [...] will leave from its imaginary moving some vestige in length, exempt of any width. [...] The straight line is that which is equally extended between its points"

Euclid described a line as "breadthless length", and introduced several postulates as basic unprovable properties from which he constructed the geometry, which is now called Euclidean geometry to avoid confusion with other geometries which have been introduced since the end of nineteenth century (such as non-Euclidean, projective and affine geometry).

In modern mathematics, given the multitude of geometries, the concept of a line is closely tied to the way the geometry is described. For instance, in analytic geometry, a line in the plane is often defined as the set of points whose coordinates satisfy a given linear equation, but in a more abstract setting, such as incidence geometry, a line may be an independent object, distinct from the set of points which lie on it.

When a geometry is described by a set of axioms, the notion of a line is usually left undefined (a so-called primitive object). The properties of lines are then determined by the axioms which refer to them. One advantage to this approach is the flexibility it gives to users of the geometry. Thus in differential geometry a line may be interpreted as a geodesic (shortest path between points), while in some projective geometries a line is a 2-dimensional vector space (all linear combinations of two independent vectors). This flexibility also extends beyond mathematics and, for example, permits physicists to think of the path of a light ray as being a line.

A line segment is a part of a line that is bounded by two distinct end points and contains every point on the line between its end points. Depending on how the line segment is defined, either of the two end points may or may not be part of the line segment. Two or more line segments may have some of the same relationships as lines, such as being parallel, intersecting, or skew.

Definitions versus Descriptions

All definitions are ultimately circular in nature since they depend on concepts which must themselves have definitions, a dependence which can not be continued indefinitely without returning to the starting point. To avoid this vicious circle certain concepts must be taken as primitive concepts; terms which are given no definition. In geometry, it is frequently the case that the concept of line is taken as a primitive. In those situations where a line is a defined concept, as in coordinate geometry, some other fundamental ideas are taken as primitives. When the line concept is a primitive, the behaviour and properties of lines are dictated by the axioms which they must satisfy.

In a non-axiomatic or simplified axiomatic treatment of geometry, the concept of a primitive notion may be too abstract to be dealt with. In this circumstance it is possible that a *description* or *mental image* of a primitive notion is provided to give a foundation to build the notion on which would formally be based on the (unstated) axioms. Descriptions of this type may be referred to, by some authors, as definitions in this informal style of presentation. These are not true definitions and could not be used in formal proofs of statements. The "definition" of line in Euclid's Elements falls into this category. Even in the case where a specific geometry is being considered (for example, Euclidean geometry), there is no generally accepted agreement among authors as to what an informal description of a line should be when the subject is not being treated formally.

Ray

Given a line and any point A on it, we may consider A as decomposing this line into two parts. Each such part is called a ray (or half-line) and the point A is called its *initial point*. The point A is considered to be a member of the ray. Intuitively, a ray consists of those points on a line passing through A and proceeding indefinitely, starting at A, in one direction only along the line. However, in order to use this concept of a ray in proofs a more precise definition is required.

Given distinct points A and B, they determine a unique ray with initial point A. As two points define a unique line, this ray consists of all the points between A and B (including A and B) and all the points C on the line through A and B such that B is between A and C. This is, at times, also expressed as the set of all points C such that A is not between B and C. A point D, on the line determined by A and B but not in the ray with initial point A determined by B, will determine another ray with initial point A. With respect to the AB ray, the AD ray is called the *opposite ray*.

Thus, we would say that two different points, A and B, define a line and a decomposition of this line into the disjoint union of an open segment (A, B) and two rays, BC and AD (the point D is not drawn in the diagram, but is to the left of A on the line AB). These are not opposite rays since they have different initial points.

In Euclidean geometry two rays with a common endpoint form an angle.

The definition of a ray depends upon the notion of betweenness for points on a line. It follows that rays exist only for geometries for which this notion exists, typically Euclidean geometry or affine geometry over an ordered field. On the other hand, rays do not exist in projective geometry nor in a geometry over a non-ordered field, like the complex numbers or any finite field.

In topology, a ray in a space X is a continuous embedding $R^+ \rightarrow X$. It is used to define the important concept of end of the space.

Euclidean Geometry

When geometry was first formalised by Euclid in the Elements, he defined a line to be "breadthless length" with a straight line being a line "which lies evenly with the points on itself". These definitions serve little purpose since they use terms which are not, themselves, defined. In fact, Euclid did not use these definitions in this work and probably included them just to make it clear to the reader what was being discussed. In modern geometry, a line is simply taken as an undefined object with properties given by axioms, but is sometimes defined as a set of points obeying a linear relationship when some other fundamental concept is left undefined.

In an axiomatic formulation of Euclidean geometry, such as that of Hilbert (Euclid's original axioms contained various flaws which have been corrected by modern mathematicians), a line is stated to have certain properties which relate it to other lines and points. For example, for any two distinct points, there is a unique line containing them, and any two distinct lines intersect in at most one point. In two dimensions, i.e., the Euclidean plane, two lines which do not intersect are called parallel. In higher dimensions, two lines that do not intersect may be parallel if they are contained in a plane, or skew if they are not.

Any collection of finitely many lines partitions the plane into convex polygons (possibly unbounded); this partition is known as an arrangement of lines.

Cartesian Plane

Lines in a Cartesian plane or, more generally, in affine coordinates, can be described algebraically by *linear* equations. In two dimensions, the equation for non-vertical lines is often given in the slope-intercept form:

$$y = mx + b$$

where:

m is the slope or gradient of the line.

b is the y-intercept of the line.

x is the independent variable of the function $y = f(x)$.

The slope of the line through points $A(x_a, y_a)$ and $B(x_b, y_b)$, when x_a 'd x_b, is given by $m = (y_b - y_a)/(x_b - x_a)$ and the equation of this line can be written y = $m(x$ d $x_a) + y_a$.

In R^2, every line L (including vertical lines) is described by a linear equation of the form

$$L = \{(x, y) \mid ax + by = c\}$$

with fixed real coefficients a, b and c such that a and b are not both zero. Using this form, vertical lines correspond to the equations with $b = 0$.

There are many variant ways to write the equation of a line which can all be converted from one to another by algebraic manipulation. These forms are generally named by the type of information (data) about the line that is needed to write down the form. Some of the important data of a line is its slope, x-intercept, known points on the line and y-intercept.

The equation of the line passing through two different points $P_0 = (x_0, y_0)$ and $P_1 = (x_1, y_1)$ may be written as

$$(y - y_0)(x_1 - x_0) = (y_1 - y_0)(x - x_0).$$

If $x_0 \neq x_1$, this equation may be rewritten as

$$y = (x - x_0)\frac{y_1 - y_0}{x_1 - x_0} + y_0$$

or

$$y = x\frac{y_1 - y_0}{x_1 - x_0} + \frac{x_1 y_0 - x_0 y_1}{x_1 - x_0}.$$

In three dimensions, lines can *not* be described by a single linear equation, so they are frequently described by parametric equations:

$$x = x_0 + at$$

$$y = y_0 + bt$$

$$z = z_0 + ct$$

where:

x, y, and z are all functions of the independent variable t which ranges over the real numbers.

(x_0, y_0, z_0) is any point on the line.

a, b, and c are related to the slope of the line, such that the vector (a, b, c) is parallel to the line.

They may also be described as the simultaneous solutions of two linear equations

$$a_1x + b_1y + c_1z - d_1 = 0$$

$$a_2x + b_2y + c_2z - d_2 = 0$$

such that (a_1, b_1, c_1) and (a_2, b_2, c_2) are not proportional (the relations $a_1 = ta_2, b_1 = tb_2, c_1 = tc_2$ imply $t = 0$). This follows since in three dimensions a single linear equation typically describes a plane and a line is what is common to two distinct intersecting planes.

Normal Form

The normal *segment* for a given line is defined to be the line segment drawn from the origin perpendicular to the line. This segment joins the origin with the closest point on the line to the origin. The *normal form* of the equation of a straight line on the plane is given by:

$$y \sin\theta + x \cos\theta - p = 0,$$

where θ is the angle of inclination of the normal segment (the oriented angle from the unit vector of the x axis to this segment), and p is the (positive) length of the normal segment. The normal form can be derived from the general form by dividing all of the coefficients by

$$\frac{|c|}{-c}\sqrt{a^2 + b^2}.$$

This form is also called the Hesse normal form,[12] after the German mathematician Ludwig Otto Hesse.

Unlike the slope-intercept and intercept forms, this form can represent any line but also requires only two finite parameters, θ and p, to be specified. Note that if $p > 0$, then θ is uniquely defined modulo 2π. On the other hand, if the line is through the origin ($c = 0$, $p = 0$), one drops the $|c|/(dc)$ term to compute sinθ and cosθ, and θ is only defined modulo π.

Polar Coordinates

In polar coordinates on the Euclidean plane a line is expressed as

$$r = \frac{mr\cos\theta + b}{\sin\theta},$$

where m is the slope of the line and b is the y-intercept. When $\theta = 0$ the graph will be undefined. The equation can be rewritten to eliminate discontinuities:

$$r\sin\theta = mr\cos\theta + b.$$

Vector Equation

The vector equation of the line through points A and B is given by r = OA + λAB (where λ is a scalar).

If a is vector OA and b is vector OB, then the equation of the line can be written: r = a + λ(b d a).

A ray starting at point A is described by limiting λ. One ray is obtained if λ ed 0, and the opposite ray comes from λ dd 0.

Euclidean Space

In three-dimensional space, a first degree equation in the variables x, y, and z defines a plane, so two such equations, provided the planes they give rise to are not parallel, define a line which is the intersection of the planes. More generally, in n-dimensional space n-1 first-degree equations in the n coordinate variables define a line under suitable conditions.

In more general Euclidean space, R^n (and analogously in every other affine space), the line L passing through two different points a and b (considered as vectors) is the subset

$$L = \{(1-t)a + tb \mid t \in \mathbb{R}\}$$

The direction of the line is from a ($t = 0$) to b ($t = 1$), or in other words, in the direction of the vector b ” a. Different choices of a and b can yield the same line.

Collinear Points

Three points are said to be *collinear* if they lie on the same line. Three points usually determine a plane, but in the case of three collinear points this does *not* happen.

In affine coordinates, in n-dimensional space the points $X=(x_1, x_2, \ldots, x_n)$, $Y=(y_1, y_2, \ldots, y_n)$, and $Z=(z_1, z_2, \ldots, z_n)$ are collinear if the matrix

$$\begin{bmatrix} 1 & x_1 & x_2 & \cdots & x_n \\ 1 & y_1 & y_2 & \cdots & y_n \\ 1 & z_1 & z_2 & \cdots & z_n \end{bmatrix}$$

has a rank less than 3. In particular, for three points in the plane ($n = 2$), above matrix is square and the points are collinear if and only if its determinant is zero.

In Euclidean geometry, the Euclidean distance $d(a,b)$ between two points a and b may be used to express the collinearity between three points by:

> *The points* a, b *and* c *are collinear if and only if* d(x,a) = d(c,a) *and* d(x,b) = d(c,b) *implies* x=c.

However there are other notions of distance (such as the Manhattan distance) for which this property is not true.

In the geometries where the concept of a line is a primitive notion, as may be the case in some synthetic geometries, other methods of determining collinearity are needed.

Types of Lines

In a sense,[15] all lines in Euclidean geometry are equal, in that, without coordinates, one can not tell them apart from one another. However, lines may play special roles with respect to other objects in the geometry and be divided into types according to that relationship. For instance, with respect to a conic (a circle, ellipse, parabola, or hyperbola), lines can be:

- tangent lines, which touch the conic at a single point;
- secant lines, which intersect the conic at two points and pass through its interior;
- exterior lines, which do not meet the conic at any point of the Euclidean plane; or
- a directrix, whose distance from a point helps to establish whether the point is on the conic.

In the context of determining parallelism in Euclidean geometry, a transversal is a line that intersects two other lines that may or not be parallel to each other.

For more general algebraic curves, lines could also be:

- *i*-secant lines, meeting the curve in *i* points counted without multiplicity, or
- asymptotes, which a curve approaches arbitrarily closely without touching it.

With respect to triangles we have:

- the Euler line, and
- the Simson lines.

For a hexagon with vertices lying on a conic we have the Pascal line and, in the special case where the conic is a pair of lines, we have the Pappus line.

Parallel lines are lines in the same plane that never cross. Intersecting lines share a single point in common. Coincidental lines coincide with each other—every point that is on either one of them is also on the other.

In three-dimensional space, skew lines are lines that are not in the same plane and thus do not intersect each other.

Projective Geometry

In many models of projective geometry, the representation of a line rarely conforms to the notion of the "straight curve" as it is visualised in Euclidean geometry. In Elliptic geometry we see a typical example of this.[16] In the spherical representation of elliptic geometry, lines are represented by great circles of a sphere with diametrically opposite points identified. In a different model of elliptic geometry, lines are represented by Euclidean planes passing through the origin. Even though these representations are visually distinct, they satisfy all the properties (such as, two points determining a unique line) that make them suitable representations for lines in this geometry.

Geodesics

The "straightness" of a line, interpreted as the property that the distance along the line between any two of its points is minimized, can be generalized and leads to the concept of geodesics in metric spaces.

Chapter 5

Matrices

In mathematics, a matrix (plural matrices) is a rectangular array of numbers, symbols, or expressions, arranged in rows and columns. The individual items in a matrix are called its *elements* or *entries*. An example of a matrix with 2 rows and 3 columns is

$$\begin{bmatrix} 1 & 9 & -13 \\ 20 & 5 & -6 \end{bmatrix}.$$

Matrices of the same size can be added or subtracted element by element. The rule for matrix multiplication, however, is that two matrices can be multiplied only when the number of columns in the first equals the number of rows in the second. A major application of matrices is to represent linear transformations, that is, generalizations of linear functions such as $f(x) = 4x$. For example, the rotation of vectors in three dimensional space is a linear transformation which can be represented by a rotation matrix R. If v is a column vector (a matrix with only one column) describing the position of a point in space, the product Rv is a column vector describing the position of that point after a rotation. The product of two matrices is a matrix that represents the composition of two linear transformations. Another application of matrices is in the solution of a system of linear equations. If the matrix is square, it is possible to deduce some of its properties by computing its determinant. For example, a square matrix has an inverse if and only if its determinant is not zero. Eigenvalues and eigenvectors provide insight into the geometry of linear transformations.

Applications of matrices are found in most scientific fields. In every branch of physics, including classical mechanics, optics, electromagnetism, quantum mechanics, and quantum electrodynamics, they are used to study physical phenomena, such as the motion of rigid bodies. In computer graphics, they are used to project a 3-dimensional image onto a 2-dimensional screen. In probability theory and statistics, stochastic matrices are used to describe sets of probabilities; for instance, they are used within the PageRank algorithm that ranks the pages in a Google search. Matrix calculus generalizes classical analytical notions such as derivatives and exponentials to higher dimensions.

A major branch of numerical analysis is devoted to the development of efficient algorithms for matrix computations, a subject that is centuries old and is today an expanding area of research. Matrix decomposition methods simplify computations, both theoretically and practically. Algorithms that are tailored to particular matrix structures, such as sparse matrices and near-diagonal matrices, expedite computations in finite element method and other computations. Infinite matrices occur in planetary theory and in atomic theory. A simple example of an infinite matrix is the matrix representing the derivative operator, which acts on the Taylor series of a function.

Definition

A *matrix* is a rectangular array of numbers or other mathematical objects, for which operations such as addition and multiplication are defined. Most commonly, a matrix over a field F is a rectangular array of scalars from F. Most of this article focuses on *real* and *complex matrices*, i.e., matrices whose elements are real numbers or complex numbers, respectively. More general types of entries are discussed below. For instance, this is a real matrix:

$$\mathbf{A} = \begin{bmatrix} -1.3 & 0.6 \\ 20.4 & 5.5 \\ 9.7 & -6.2 \end{bmatrix}.$$

The numbers, symbols or expressions in the matrix are called its *entries* or its *elements*. The horizontal and vertical lines of entries in a matrix are called *rows* and *columns*, respectively.

Size

The size of a matrix is defined by the number of rows and columns that it contains. A matrix with m rows and n columns is called an $m \times n$ matrix or m-by-n matrix, while m and n are called its *dimensions*. For example, the matrix A above is a 3×2 matrix.

Matrices which have a single row are called row vectors, and those which have a single column are called column vectors. A matrix which has the same number of rows and columns is called a square matrix. A matrix with an infinite number of rows or columns (or both) is called an infinite matrix. In some contexts, such as computer algebra programs, it is useful to consider a matrix with no rows or no columns, called an empty matrix.

Name	*Size*	*Example Description*
Row vector	$1 \times n$	A matrix with one row, sometimes used to represent a vector
Column vector	$n \times 1$	A matrix with one column, sometimes used to represent a vector
Square matrix	$n \times n$	A matrix with the same number of rows and columns, sometimes used to represent a linear transformation from a vector space to itself, such as reflection, rotation, or shearing.

Notation

Matrices are commonly written in box brackets:

$$\mathbf{A} = \begin{bmatrix} a_{11} & a_{12} & \cdots & a_{1n} \\ a_{21} & a_{22} & \cdots & a_{2n} \\ \vdots & \vdots & \ddots & \vdots \\ a_{m1} & a_{m2} & \cdots & a_{mn} \end{bmatrix}.$$

An alternative notation uses large parentheses instead of box brackets:

$$\mathbf{A} = \begin{pmatrix} a_{11} & a_{12} & \cdots & a_{1n} \\ a_{21} & a_{22} & \cdots & a_{2n} \\ \vdots & \vdots & \ddots & \vdots \\ a_{m1} & a_{m2} & \cdots & a_{mn} \end{pmatrix}.$$

The specifics of symbolic matrix notation varies widely, with some prevailing trends. Matrices are usually symbolized using upper-case letters (such as A in the examples above), while the corresponding lower-case letters, with two subscript indices (e.g., a_{11}, or $a_{1,1}$), represent the entries. In addition to using upper-case letters to symbolize matrices, many authors use a special typographical style, commonly boldface upright (non-italic), to further distinguish matrices from other mathematical objects. An alternative notation involves the use of a double-underline with the variable name, with or without boldface style, (e.g., $\underline{\underline{A}}$).

The entry in the i-th row and j-th column of a matrix A is sometimes referred to as the i,j, (i,j), or $(i,j)^{\text{th}}$ entry of the matrix, and most commonly denoted as $a_{i,j}$, or a_{ij}. Alternative notations for that entry are $A[i,j]$ or $A_{i,j}$. For example, the (1,3) entry of the following matrix A is 5 (also denoted a_{13}, $a_{1,3}$, $A[1,3]$ or $A_{1,3}$):

$$\mathbf{A} = \begin{bmatrix} 4 & -7 & 5 & 0 \\ -2 & 0 & 11 & 8 \\ 19 & 1 & -3 & 12 \end{bmatrix}$$

Sometimes, the entries of a matrix can be defined by a formula such as $a_{i,j} = f(i, j)$. For example, each of the entries of the following matrix A is determined by $a_{ij} = i - j$.

$$\mathbf{A} = \begin{bmatrix} 0 & -1 & -2 & -3 \\ 1 & 0 & -1 & -2 \\ 2 & 1 & 0 & -1 \end{bmatrix}$$

In this case, the matrix itself is sometimes defined by that formula, within square brackets or double parenthesis. For example, the matrix above is defined as A = $[i\text{-}j]$, or A = $((i\text{-}j))$. If matrix size is $m \times n$, the above-mentioned formula $f(i, j)$ is valid for any $i = 1, ..., m$ and any $j = 1, ..., n$. This can be either specified separately, or using $m \times n$ as a subscript. For instance, the matrix A above is 3×4 and can be defined as A = $[i - j]$ ($i = 1, 2, 3; j = 1, ..., 4$), or A = $[i - j]_{3\times 4}$.

Some programming languages utilize doubly subscripted arrays (or arrays of arrays) to represent an m-×-n matrix. Some programming languages start the numbering of array indexes at zero, in which case the entries of an m-by-n matrix are indexed by $0 \le i \le m - 1$ and $0 \le j \le n - 1$. This article follows the more common convention in mathematical writing where enumeration starts from 1.

Operation	*Definition*	*Example*
Addition	The *sum* A+B of two m-by-n matrices A and B is calculated entrywise: $(A + B)_{i,j} = A_{i,j} + B_{i,j}$, where $1 \le i \le m$ and $1 \le j \le n$.	$\begin{bmatrix} 1 & 3 & 1 \\ 1 & 0 & 0 \end{bmatrix} + \begin{bmatrix} 0 & 0 & 5 \\ 7 & 5 & 0 \end{bmatrix} = \begin{bmatrix} 1+0 & 3+0 & 1+5 \\ 1+7 & 0+5 & 0+0 \end{bmatrix} = \begin{bmatrix} 1 & 3 & 6 \\ 8 & 5 & 0 \end{bmatrix}$
Scalar multiplication	The *scalar multiplication* cA of a matrix A and a number c (also called a scalar in the parlance of abstract algebra) is given by multiplying every entry of A by c: $(cA)_{i,j} = c \cdot A_{i,j}$.	$2 \cdot \begin{bmatrix} 1 & 8 & -3 \\ 4 & -2 & 5 \end{bmatrix} = \begin{bmatrix} 2\cdot 1 & 2\cdot 8 & 2\cdot -3 \\ 2\cdot 4 & 2\cdot -2 & 2\cdot 5 \end{bmatrix} = \begin{bmatrix} 2 & 16 & -6 \\ 8 & -4 & 10 \end{bmatrix}$
Transpose	The *transpose* of an m-by-n matrix A is the n-by-m matrix A^T (also denoted A^{tr} or ${}^tA$) formed by turning rows into columns and vice versa: $(A^T)_{i,j} = A_{j,i}$.	$\begin{bmatrix} 1 & 2 & 3 \\ 0 & -6 & 7 \end{bmatrix}^T = \begin{bmatrix} 1 & 0 \\ 2 & -6 \\ 3 & 7 \end{bmatrix}$

Basic Operations

There are a number of basic operations that can be applied to modify matrices, called *matrix addition, scalar multiplication, transposition, matrix multiplication, row operations*, and *submatrix.*

Addition, Scalar Multiplication and Transposition

Familiar properties of numbers extend to these operations of matrices: for example, addition is commutative, i.e., the matrix sum does not depend on the order of the summands: A + B = B + A. The transpose is compatible with addition and scalar multiplication, as expressed by $(cA)^T = c(A^T)$ and $(A + B)^T = A^T + B^T$. Finally, $(A^T)^T = A$.

Matrix Multiplication

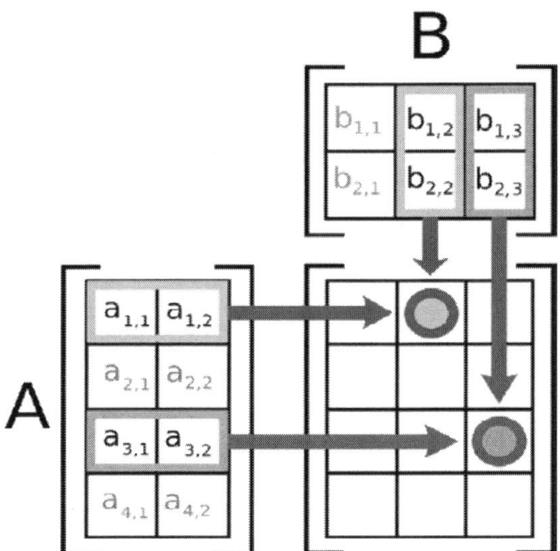

Figure: *Schematic depiction of the matrix product AB of two matrices A and B.*

Multiplication of two matrices is defined if and only if the number of columns of the left matrix is the same as the number of rows of the right matrix. If A is an m-by-n matrix and B is an n-by-p matrix, then their *matrix product* AB is the m-by-p matrix whose entries are given by dot product of the corresponding row of A and the corresponding column of B:

$$[\mathbf{AB}]_{i,j} = A_{i,1}B_{1,j} + A_{i,2}B_{2,j} + \cdots + A_{i,n}B_{n,j} = \sum_{r=1}^{n} A_{i,r}B_{r,j},$$

where $1 \le i \le m$ and $1 \le j \le p$. For example, the underlined entry 2340 in the product is calculated as $(2 \times 1000) + (3 \times 100) + (4 \times 10) = 2340$:

$$\begin{bmatrix} \underline{2} & \underline{3} & \underline{4} \\ 1 & 0 & 0 \end{bmatrix} \begin{bmatrix} 0 & \underline{1000} \\ 1 & \underline{100} \\ 0 & \underline{10} \end{bmatrix} = \begin{bmatrix} 3 & \underline{2340} \\ 0 & 1000 \end{bmatrix}.$$

Matrix multiplication satisfies the rules (AB)C = A(BC) (associativity), and (A+B)C = AC+BC as well as C(A+B) = CA+CB (left and right distributivity), whenever the size of the matrices is such that the various products are defined.[13] The product AB may be defined without BA being defined, namely if A and B are m-by-n and n-by-k matrices, respectively, and $m \neq k$. Even if both products are defined, they need not be equal, i.e., generally

$$AB \neq BA,$$

i.e., *matrix multiplication is not* commutative, in marked contrast to (rational, real, or complex) numbers whose product is independent of the order of the factors. An example of two matrices not commuting with each other is:

$$\begin{bmatrix} 1 & 2 \\ 3 & 4 \end{bmatrix} \begin{bmatrix} 0 & 1 \\ 0 & 0 \end{bmatrix} = \begin{bmatrix} 0 & 1 \\ 0 & 3 \end{bmatrix},$$

whereas

$$\begin{bmatrix} 0 & 1 \\ 0 & 0 \end{bmatrix} \begin{bmatrix} 1 & 2 \\ 3 & 4 \end{bmatrix} = \begin{bmatrix} 3 & 4 \\ 0 & 0 \end{bmatrix}.$$

Besides the ordinary matrix multiplication just described, there exist other less frequently used operations on matrices that can be considered forms of multiplication, such as the Hadamard product and the Kronecker product.[14] They arise in solving matrix equations such as the Sylvester equation.

Row Operations

There are three types of row operations:

1. row addition, that is adding a row to another.
2. row multiplication, that is multiplying all entries of a row by a non-zero constant;
3. row switching, that is interchanging two rows of a matrix;

These operations are used in a number of ways, including solving linear equations and finding matrix inverses.

Submatrix

A submatrix of a matrix is obtained by deleting any collection of rows and/or columns. For example, for the following 3-by-4 matrix, we can construct a 2-by-3 submatrix by removing row 3 and column 2:

$$\mathbf{A} = \begin{bmatrix} 1 & 2 & 3 & 4 \\ 5 & 6 & 7 & 8 \\ 9 & 10 & 11 & 12 \end{bmatrix} \rightarrow \begin{bmatrix} 1 & 3 & 4 \\ 5 & 7 & 8 \end{bmatrix}.$$

The minors and cofactors of a matrix are found by computing the determinant of certain submatrices.

Linear Equations

Matrices can be used to compactly write and work with multiple linear equations, i.e., systems of linear equations. For example, if A is an m-by-n matrix, x designates a column vector (i.e., $n\times1$-matrix) of n variables x_1, x_2, ..., x_n, and b is an $m\times1$-column vector, then the matrix equation

$$Ax = b$$

is equivalent to the system of linear equations

$$A_{1,1}x_1 + A_{1,2}x_2 + \ldots + A_{1,n}x_n = b_1$$

$$\ldots$$

$$A_{m,1}x_1 + A_{m,2}x_2 + \ldots + A_{m,n}x_n = b_m\ .$$

Linear Transformations

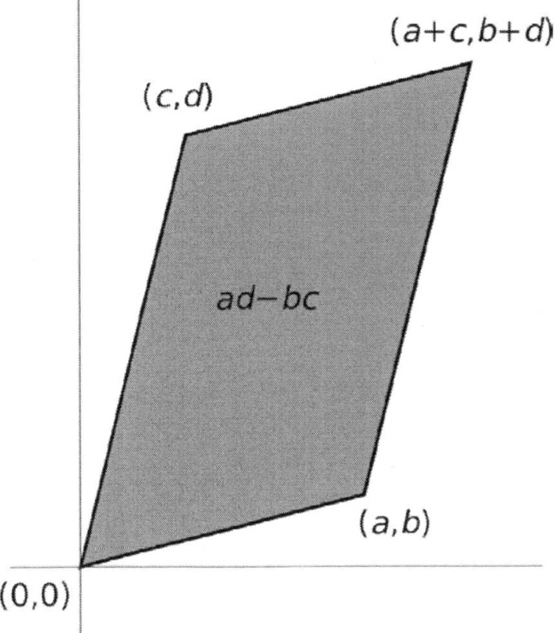

Figure: *The vectors represented by a 2-by-2 matrix correspond to the sides of a unit square transformed into a parallelogram.*

Matrices and matrix multiplication reveal their essential features when related to *linear transformations*, also known as *linear maps. A real m-by-n matrix A gives rise to a linear transformation* $R^n \to R^m$ *mapping each vector x in* R^n *to the (matrix) product Ax, which is a vector in* R^m*. Conversely, each linear transformation* $f: R^n \to R^m$ *arises from a unique m-by-n matrix A: explicitly, the (i, j)-entry of A is the* i^{th} *coordinate of* $f(e_j)$*, where* $e_j = (0,\ldots,0,1,0,\ldots,0)$ *is the* unit vector *with 1 in the* j^{th} *position and 0 elsewhere.* The matrix A is said to represent the linear map *f*, and A is called the *transformation matrix* of *f*.

Square Matrices

A square matrix is a matrix with the same number of rows and columns. An *n*-by-*n* matrix is known as a square matrix of order *n*. Any two square matrices of the same order can be added and multiplied. The entries a_{ii} form the main diagonal of a square matrix. They lie on the imaginary line which runs from the top left corner to the bottom right corner of the matrix.

Main Types

Name	*Example with n = 3*
Diagonal matrix	$\begin{bmatrix} a_{11} & 0 & 0 \\ 0 & a_{22} & 0 \\ 0 & 0 & a_{33} \end{bmatrix}$
Lower triangular matrix	$\begin{bmatrix} a_{11} & 0 & 0 \\ a_{21} & a_{22} & 0 \\ a_{31} & a_{32} & a_{33} \end{bmatrix}$
Upper triangular matrix	$\begin{bmatrix} a_{11} & a_{12} & a_{13} \\ 0 & a_{22} & a_{23} \\ 0 & 0 & a_{33} \end{bmatrix}$

Diagonal and Triangular Matrices

If all entries of A below the main diagonal are zero, A is called an *upper* triangular matrix. Similarly if all entries of *A* above the main diagonal are zero, A is called a *lower triangular matrix.* If all entries outside the main diagonal are zero, A is called a diagonal matrix.

Identity Matrix

The identity matrix I_n of size n is the n-by-n matrix in which all the elements on the main diagonal are equal to 1 and all other elements are equal to 0, e.g.

$$I_1 = [1],\ I_2 = \begin{bmatrix} 1 & 0 \\ 0 & 1 \end{bmatrix}, \cdots, I_n = \begin{bmatrix} 1 & 0 & \cdots & 0 \\ 0 & 1 & \cdots & 0 \\ \vdots & \vdots & \ddots & \vdots \\ 0 & 0 & \cdots & 1 \end{bmatrix}$$

It is a square matrix of order n, and also a special kind of diagonal matrix. It is called identity matrix because multiplication with it leaves a matrix unchanged:

$AI_n = I_mA = A$ for any m-by-n matrix A.

Symmetric or Skew-Symmetric Matrix

A square matrix A that is equal to its transpose, i.e., $A = A^T$, is a symmetric matrix. If instead, A was equal to the negative of its transpose, i.e., $A = -A^T$, then A is a skew-symmetric matrix. In complex matrices, symmetry is often replaced by the concept of Hermitian matrices, which satisfy $A^- = A$, where the star or asterisk denotes the conjugate transpose of the matrix, i.e., the transpose of the complex conjugate of A.

By the spectral theorem, real symmetric matrices and complex Hermitian matrices have an eigenbasis; i.e., every vector is expressible as a linear combination of eigenvectors. In both cases, all eigenvalues are real.

This theorem can be generalized to infinite-dimensional situations related to matrices with infinitely many rows and columns.

Invertible Matrix and its Inverse

A square matrix A is called invertible or *non-singular* if there exists a matrix B such that

$$AB = BA = I_n.$$

If B exists, it is unique and is called the inverse matrix of A, denoted A^{-1}.

Definite Matrix

Positive definite matrix	*Indefinite matrix*
$\begin{bmatrix} 1/4 & 0 \\ 0 & 1 \end{bmatrix}$	$\begin{bmatrix} 1/4 & 0 \\ 0 & -1/4 \end{bmatrix}$
$Q(x,y) = 1/4x^2 + y^2$	$Q(x,y) = 1/4x^2 - 1/4y^2$

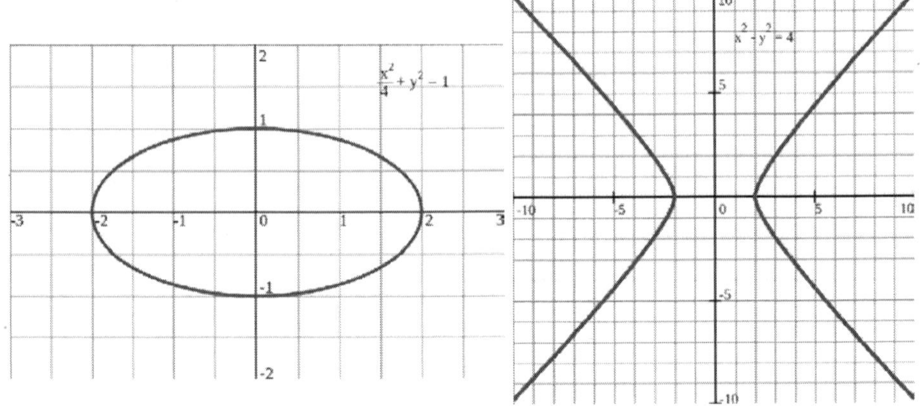

Points such that Q(x,y)=1 (Ellipse) *Points such that Q(x,y)=1 (Hyperbola).*

A symmetric $n{\times}n$-matrix is called positive-definite (respectively negative-definite; indefinite), if for all nonzero vectors x – R^n the associated quadratic form given by

$$Q(x) = x^TAx$$

takes only positive values (respectively only negative values; both some negative and some positive values). If the quadratic form takes only non-negative (respectively only non-positive) values, the symmetric matrix is called positive-semidefinite (respectively negative-semidefinite); hence the matrix is indefinite precisely when it is neither positive-semidefinite nor negative-semidefinite. A symmetric matrix is positive-definite if and only if all its eigenvalues are positive, i.e., the matrix is positive-semidefinite and it is invertible. The table at the right shows two possibilities for 2-by-2 matrices. Allowing as input two different vectors instead yields the bilinear form associated to A:

$$B_A(x, y) = x^TAy.$$

Orthogonal Matrix

An *orthogonal matrix* is a square matrix with real entries whose columns and rows are orthogonal unit vectors (i.e., orthonormal vectors). Equivalently, a matrix A is orthogonal if its transpose is equal to its inverse:

$$A^{\mathrm{T}} = A^{-1},$$

which entails

$$A^{\mathrm{T}} A = AA^{\mathrm{T}} = I,$$

where I is the identity matrix.

An orthogonal matrix A is necessarily invertible (with inverse $A^{-1} = A^{\mathrm{T}}$), unitary ($A^{-1} = A^*$), and normal ($A^*A = AA^*$). The determinant of any orthogonal matrix is either +1 or −1. A *special orthogonal matrix* is an orthogonal matrix with determinant +1. As a linear transformation, every orthogonal matrix with determinant +1 is a pure rotation, while every orthogonal matrix with determinant -1 is either a pure reflection, or a composition of reflection and rotation.

The complex analogue of an orthogonal matrix is a unitary matrix.

Main Operations

Trace: The trace, tr(A) of a square matrix A is the sum of its diagonal entries. While matrix multiplication is not commutative as mentioned above, the trace of the product of two matrices is independent of the order of the factors:

$$\mathrm{tr(AB)} = \mathrm{tr(BA)}.$$

This is immediate from the definition of matrix multiplication:

$$\mathrm{tr}(\mathsf{AB}) = \sum_{i=1}^{m}\sum_{j=1}^{n} A_{ij}B_{ji} = \mathrm{tr}(\mathsf{BA}).$$

Also, the trace of a matrix is equal to that of its transpose, i.e.,

$$\mathrm{tr(A)} = \mathrm{tr(A^T)}.$$

Determinant

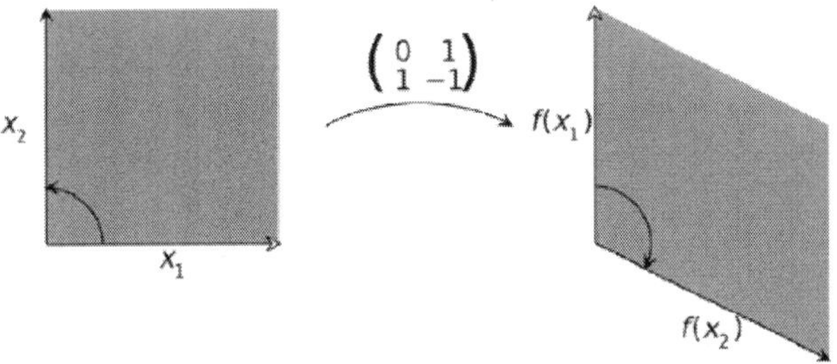

Figure: *A linear transformation on R^2 given by the indicated matrix. The determinant of this matrix is –1, as the area of the green parallelogram at the right is 1, but the map reverses the orientation, since it turns the counterclockwise orientation of the vectors to a clockwise one.*

The *determinant* det(A) or |A| of a square matrix A is a number encoding certain properties of the matrix. A matrix is invertible if and only if its determinant is nonzero. Its absolute value equals the area (in R^2) or volume (in R^3) of the image of the unit square (or cube), while its sign corresponds to the orientation of the corresponding linear map: the determinant is positive if and only if the orientation is preserved.

The determinant of 2-by-2 matrices is given by

$$\det\begin{bmatrix} a & b \\ c & d \end{bmatrix} = ad - bc.$$

The determinant of 3-by-3 matrices involves 6 terms (rule of Sarrus). The more lengthy Leibniz formula generalises these two formulae to all dimensions.[26]

The determinant of a product of square matrices equals the product of their determinants:

$$\det(AB) = \det(A) \cdot \det(B).$$

Adding a multiple of any row to another row, or a multiple of any column to another column, does not change the determinant. Interchanging two rows or two columns affects the determinant by multiplying it by –1. Using these operations, any matrix can be transformed to a lower (or upper) triangular matrix, and for such matrices the determinant equals the product of the entries on the main diagonal; this provides a method to calculate the determinant of any matrix. Finally, the Laplace expansion expresses the determinant in terms of minors, i.e., determinants of smaller matrices. This expansion can be used for a recursive definition of determinants (taking as starting case the determinant of a 1-by-1 matrix, which is its unique entry, or even the determinant of a 0-by-0 matrix, which is 1), that can be seen to be equivalent to the Leibniz formula. Determinants can be used to solve linear systems using Cramer's rule, where the division of the determinants of two related square matrices equates to the value of each of the system's variables.

Eigenvalues and Eigenvectors

A number λ and a non-zero vector v satisfying

$$Av = \lambda v$$

are called an *eigenvalue* and an *eigenvector* of A, respectively. The number λ is an eigenvalue of an $n\times n$-matrix A if and only if $A-\lambda I_n$ is not invertible, which is equivalent to

$$\det(A - \lambda I) = 0.$$

The polynomial p_A in an indeterminate X given by evaluation the determinant $\det(XI_n - A)$ is called the characteristic polynomial of A. It is a monic polynomial of degree n. Therefore the polynomial equation $p_A(\lambda) = 0$ has at most n different solutions, i.e., eigenvalues of the matrix. They may be complex even if the entries of A are real. According to the Cayley–Hamilton theorem, $p_A(A) = 0$, that is, the result of substituting the matrix itself into its own characteristic polynomial yields the zero matrix.

Computational Aspects

Matrix calculations can be often performed with different techniques. Many problems can be solved by both direct algorithms or iterative approaches. For example, the eigenvectors of a square matrix can be obtained by finding a sequence of vectors x_n converging to an eigenvector when n tends to infinity.

To be able to choose the more appropriate algorithm for each specific problem, it is important to determine both the effectiveness and precision of all the available algorithms. The domain studying these matters is called numerical linear algebra. As with other numerical situations, two main aspects are the complexity of algorithms and their numerical stability.

Determining the complexity of an algorithm means finding upper bounds or estimates of how many elementary operations such as additions and multiplications of scalars are necessary to perform some algorithm, e.g., multiplication of matrices. For example, calculating the matrix product of two n-by-n matrix using the definition given above needs n^3 multiplications, since for any of the n^2 entries of the product, n multiplications are necessary. The Strassen algorithm outperforms this "naive" algorithm; it needs only $n^{2.807}$ multiplications. A refined approach also incorporates specific features of the computing devices.

In many practical situations additional information about the matrices involved is known. An important case are sparse matrices, i.e., matrices most of whose entries are zero. There are specifically adapted algorithms for, say, solving linear systems $Ax = b$ for sparse matrices A, such as the conjugate gradient method.

An algorithm is, roughly speaking, numerically stable, if little deviations in the input values do not lead to big deviations in the result. For example, calculating the inverse of a matrix via Laplace's formula (Adj (A) denotes the adjugate matrix of A)

$$A^{-1} = \mathrm{Adj}(A) / \det(A)$$

may lead to significant rounding errors if the determinant of the matrix is very small. The norm of a matrix can be used to capture the conditioning of linear algebraic problems, such as computing a matrix's inverse.

Although most computer languages are not designed with commands or libraries for matrices, as early as the 1970s, some engineering desktop computers such as the HP 9830 had ROM cartridges to add BASIC commands for matrices. Some computer languages such as APL were designed to manipulate matrices, and various mathematical programs can be used to aid computing with matrices.

Decomposition

There are several methods to render matrices into a more easily accessible form. They are generally referred to as *matrix decomposition* or *matrix factorization* techniques. The interest of all these techniques is that they preserve certain properties of the matrices in question, such as determinant, rank or inverse, so that these quantities can be calculated after applying the transformation, or that certain matrix operations are algorithmically easier to carry out for some types of matrices.

The LU decomposition factors matrices as a product of lower (L) and an upper triangular matrices (U). Once this decomposition is calculated, linear systems can be solved more efficiently, by a simple technique called forward and back substitution. Likewise, inverses of triangular matrices are algorithmically easier to calculate. The *Gaussian elimination* is a similar algorithm; it transforms any matrix to row echelon form. Both methods proceed by multiplying the matrix by suitable elementary matrices, which correspond to permuting rows or columns and adding multiples of one row to another row. Singular value decomposition expresses any matrix A as a product UDV', where U and V are unitary matrices and D is a diagonal matrix.

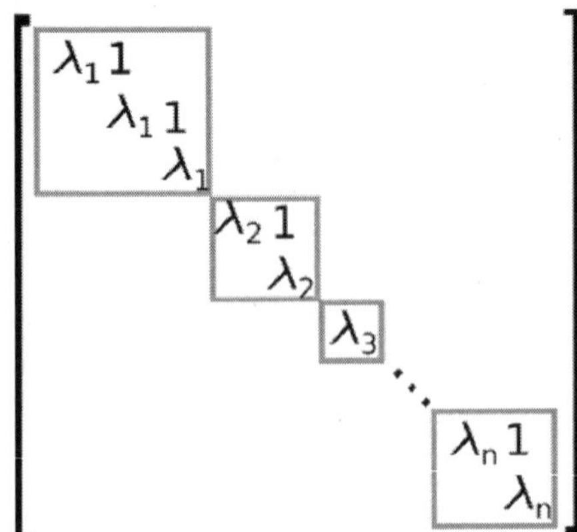

Figure: *An example of a matrix in Jordan normal form. The grey blocks are called Jordan blocks.*

The eigendecomposition or *diagonalization* expresses A as a product VDV^{-1}, where D is a diagonal matrix and V is a suitable invertible matrix. If A can be written in this form, it is called diagonalizable. More generally, and applicable to all matrices, the Jordan decomposition transforms a matrix into Jordan normal form, that is to say matrices whose only nonzero entries are the eigenvalues λ_1 to λ_n of A, placed on the main diagonal and possibly entries equal to one directly above the main diagonal, as shown at the right.[43] Given the eigendecomposition, the nth power of A (i.e., n-fold iterated matrix multiplication) can be calculated via

$$A^n = (VDV^{-1})^n = VDV^{-1}VDV^{-1}...VDV^{-1} = VD^nV^{-1}$$

and the power of a diagonal matrix can be calculated by taking the corresponding powers of the diagonal entries, which is much easier than doing the exponentiation for A instead. This can be used to compute the matrix exponential e^A, a need frequently arising in solving linear differential equations, matrix logarithms and square roots of matrices. To avoid numerically ill-conditioned situations, further algorithms such as the Schur decomposition can be employed.

Abstract Algebraic Aspects and Generalizations

Matrices can be generalized in different ways. Abstract algebra uses matrices with entries in more general fields or even rings, while linear algebra codifies properties of matrices in the notion of linear maps. It is possible to consider matrices with infinitely many columns and rows. Another extension are tensors, which can be seen as higher-dimensional arrays of numbers, as opposed to vectors, which can often be realised as sequences of numbers, while matrices are rectangular or two-dimensional array of numbers. Matrices, subject to certain requirements tend to form groups known as matrix groups.

Matrices with More General Entries

This article focuses on matrices whose entries are real or complex numbers. *However, matrices can be considered with much more general types of entries than real or complex numbers.* As a first step of generalization, any field, i.e., a set where addition, subtraction, multiplication and division operations are defined and well-behaved, may be used instead of R or C, for example rational numbers or finite fields. For example, coding theory makes use of matrices over finite fields. Wherever eigenvalues are considered, as these are roots of a polynomial they may exist only in a larger field than that of the entries of the matrix; for instance they may be complex in case of a matrix with real entries. The possibility to reinterpret the entries of a matrix as elements of a larger field (e.g., to view a real matrix as a complex matrix whose entries happen to be all real) then allows considering each square matrix to possess a full set of eigenvalues. Alternatively one can consider only matrices with entries in an algebraically closed field, such as C, from the outset.

More generally, abstract algebra makes great use of matrices with entries in a ring R. Rings are a more general notion than fields in that a division operation need not exist. The very same addition and multiplication operations of matrices extend to this setting, too. The set $M(n, R)$ of all square n-by-n matrices over R is a ring called matrix ring, isomorphic to the endomorphism ring of the left R-module R^n. If the ring R is commutative, i.e., its multiplication is commutative, then $M(n, R)$ is a unitary noncommutative (unless $n = 1$) associative algebra over R. The determinant of square matrices over a commutative ring R can still be defined using the Leibniz formula; such a matrix is invertible if and only if its determinant is invertible in R, generalising the situation over a field F, where every nonzero element is invertible.[49] Matrices over superrings are called supermatrices.

Matrices do not always have all their entries in the same ring – or even in any ring at all. One special but common case is block matrices, which may be considered as matrices whose entries themselves are matrices. The entries need not be quadratic matrices, and thus need not be members of any ordinary ring; but their sizes must fulfil certain compatibility conditions.

Relationship to Linear Maps

Linear maps $R^n \to R^m$ are equivalent to m-by-n matrices, as described above. More generally, any linear map f: $V \to W$ between finite-dimensional vector spaces can be described by a matrix $A = (a_{ij})$, after choosing bases v_1, ..., v_n of V, and w_1, ..., w_m of W (so n is the dimension of V and m is the dimension of W), which is such that

$$f(\mathbf{v}_j) = \sum_{i=1}^{m} a_{i,j}\mathbf{w}_i \qquad \text{for } j = 1,\ldots,n.$$

In other words, column j of A expresses the image of v_j in terms of the basis vectors w_i of W; thus this relation uniquely determines the entries of the matrix A. Note that the matrix depends on the choice of the bases: different choices of bases give rise to different, but equivalent matrices. Many of the above concrete notions can be reinterpreted in this light, for example, the transpose matrix A^T describes the transpose of the linear map given by A, with respect to the dual bases.

These properties can be restated in a more natural way: the category of all matrices with entries in a field k with multiplication as composition is equivalent to the category of finite dimensional vector spaces and linear maps over this field.

More generally, the set of $m\times n$ matrices can be used to represent the R-linear maps between the free modules R^m and R^n for an arbitrary ring R with unity. When $n = m$ composition of these maps is possible, and this gives rise to the matrix ring of $n\times n$ matrices representing the endomorphism ring of R^n.

Matrix Groups

A group is a mathematical structure consisting of a set of objects together with a binary operation, i.e., an operation combining any two objects to a third, subject to certain requirements. A group in which the objects are matrices and the group operation is matrix multiplication is called a *matrix group*. Since in a group every element has to be invertible, the most general matrix groups are the groups of all invertible matrices of a given size, called the general linear groups.

Any property of matrices that is preserved under matrix products and inverses can be used to define further matrix groups. For example, matrices with a given size and with a determinant of 1 form a subgroup of (i.e., a smaller group contained in) their general linear group, called a special linear group. Orthogonal matrices, determined by the condition

$$M^TM = I,$$

form the orthogonal group. Every orthogonal matrix has determinant 1 or –1. Orthogonal matrices with determinant 1 form a subgroup called *special orthogonal group*.

Every finite group is isomorphic to a matrix group, as one can see by considering the regular representation of the symmetric group. General groups can be studied using matrix groups, which are comparatively well-understood, by means of representation theory.

Infinite Matrices

It is also possible to consider matrices with infinitely many rows and/or columns even if, being infinite objects, one cannot write down such matrices explicitly. All that matters is that for every element in the set indexing rows, and every element in the set indexing columns, there is a well-defined entry (these index sets need not even be subsets of the natural numbers). The basic operations of addition, subtraction, scalar multiplication and transposition can still be defined without problem; however matrix multiplication may involve infinite summations to define the resulting entries, and these are not defined in general.

If R is any ring with unity, then the ring of endomorphisms of $M = \bigoplus_{i \in I} R$ as a right R module is isomorphic to the ring of column finite matrices $\mathbb{CFM}_I(R)$ whose entries are indexed by $I \times I$, and whose columns each contain only finitely many nonzero entries. The endomorphisms of M considered as a left R module result in an analogous object, the row finite matrices $\mathbb{RFM}_I(R)$ whose rows each only have finitely many nonzero entries.

If infinite matrices are used to describe linear maps, then only those matrices can be used all of whose columns have but a finite number of nonzero entries, for the following reason. For a matrix A to describe a linear map f: $V \to W$, bases for both spaces must have been chosen; recall that by definition this means that every vector in the space can be written uniquely as a (finite) linear combination of basis vectors, so that written as a (column) vector v of coefficients, only finitely many entries v_i are nonzero. Now the columns of A describe the images by f of individual basis vectors of V in the basis of W, which is only meaningful if these columns have only finitely many nonzero entries. There is no restriction on the rows of A however: in the product $A \cdot v$ there are only finitely many nonzero coefficients of v involved, so every one of its entries, even if it is given as an infinite sum of products, involves only finitely many nonzero terms and is therefore well defined. Moreover this amounts to forming a linear combination of the columns of A that effectively involves only finitely many of them, whence the result has only finitely many nonzero entries, because each of those columns do. One also sees that products of two matrices of the given type is well defined (provided as usual that the column-index and row-index sets match), is again of the same type, and corresponds to the composition of linear maps.

If R is a normed ring, then the condition of row or column finiteness can be relaxed. With the norm in place, absolutely convergent series can be used instead of finite sums. For example, the matrices whose column sums are absolutely convergent sequences form a ring. Analogously of course, the matrices whose row sums are absolutely convergent series also form a ring.

In that vein, infinite matrices can also be used to describe operators on Hilbert spaces, where convergence and continuity questions arise, which again results in certain constraints that have to be imposed. However, the explicit point of view of matrices tends to obfuscate the matter, and the abstract and more powerful tools of functional analysis can be used instead.

Empty Matrices

An *empty matrix* is a matrix in which the number of rows or columns (or both) is zero. Empty matrices help dealing with maps involving the zero vector space. For example, if A is a 3-by-0 matrix and B is a 0-by-3 matrix, then AB is the 3-by-3 zero matrix corresponding to the null map from a 3-dimensional space V to itself, while BA is a 0-by-0 matrix. There is no common notation for empty matrices, but most computer algebra systems allow creating and computing with them. The determinant of the 0-by-0 matrix is 1 as follows from regarding the empty product occurring in the Leibniz formula for the determinant as 1. This value is also consistent with the fact that the identity map from any finite dimensional space to itself has determinant 1, a fact that is often used as a part of the characterization of determinants.

Applications

There are numerous applications of matrices, both in mathematics and other sciences. Some of them merely take advantage of the compact representation of a set of numbers in a matrix. For example, in game theory and economics, the payoff matrix encodes the payoff for two players, depending on which out of a given (finite) set of alternatives the players choose. Text mining and automated thesaurus compilation makes use of document-term matrices such as tf-idf to track frequencies of certain words in several documents.

Complex numbers can be represented by particular real 2-by-2 matrices via

$$a+ib \leftrightarrow \begin{bmatrix} a & -b \\ b & a \end{bmatrix},$$

under which addition and multiplication of complex numbers and matrices correspond to each other. For example, 2-by-2 rotation matrices represent the multiplication with some complex number of absolute value 1, as above. A similar interpretation is possible for quaternions and Clifford algebras in general.

Early encryption techniques such as the Hill cipher also used matrices. However, due to the linear nature of matrices, these codes are comparatively easy to break. Computer graphics uses matrices both to represent objects and to calculate transformations of objects using affine rotation matrices to accomplish tasks such as projecting a three-dimensional object onto a two-dimensional screen, corresponding to a theoretical camera observation. Matrices over a polynomial ring are important in the study of control theory.

Chemistry makes use of matrices in various ways, particularly since the use of quantum theory to discuss molecular bonding and spectroscopy. Examples are the overlap matrix and the Fock matrix used in solving the Roothaan equations to obtain the molecular orbitals of the Hartree–Fock method.

Graph Theory

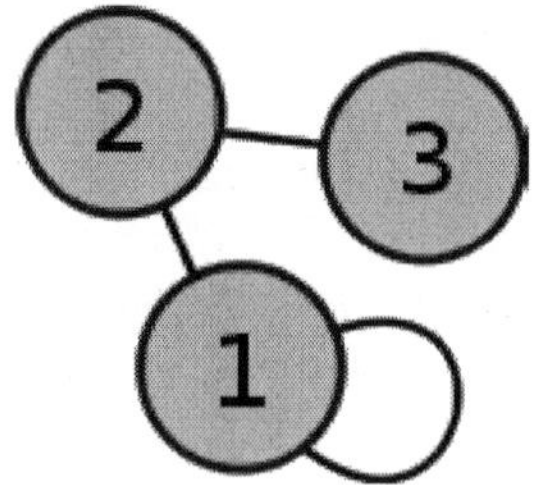

Figure: *An undirected graph with adjacency matrix* $\begin{bmatrix} 1 & 1 & 0 \\ 1 & 0 & 1 \\ 0 & 1 & 0 \end{bmatrix}$.

The adjacency matrix of a finite graph is a basic notion of graph theory. It records which vertices of the graph are connected by an edge. Matrices containing just two different values (1 and 0 meaning for example "yes" and "no", respectively) are called logical matrices. The distance (or cost) matrix contains information about distances of the edges. These concepts can be applied to websites connected hyperlinks or cities connected by roads etc., in which case (unless the road network is extremely dense) the matrices tend to be sparse, i.e., contain few nonzero entries. Therefore, specifically tailored matrix algorithms can be used in network theory.

Analysis and Geometry

The Hessian matrix of a differentiable function $f\colon \mathbb{R}^n \to \mathbb{R}$ consists of the second derivatives of f with respect to the several coordinate directions, i.e.

$$H(f) = \left[\frac{\partial^2 f}{\partial x_i \partial x_j}\right].$$

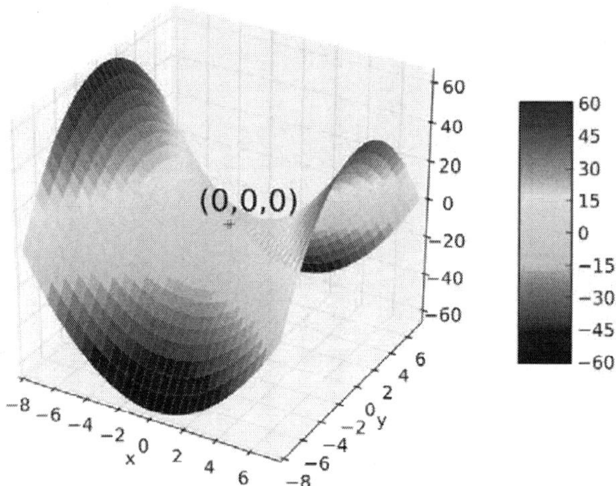

Figure: *At the saddle point* (x = *0,* y = *0) (red) of the function* f(x,–y) = $x^2 - y^2$, *the Hessian matrix* $\begin{bmatrix} 2 & 0 \\ 0 & -2 \end{bmatrix}$ *is indefinite.*

It encodes information about the local growth behaviour of the function: given a critical point x = $(x_1, ..., x_n)$, i.e., a point where the first partial derivatives $\partial f / \partial x_i$ of f vanish, the function has a local minimum if the Hessian matrix is positive definite. Quadratic programming can be used to find global minima or maxima of quadratic functions closely related to the ones attached to matrices.

Another matrix frequently used in geometrical situations is the Jacobi matrix of a differentiable map f: $R^n \to R^m$. If $f_1, ..., f_m$ denote the components of f, then the Jacobi matrix is defined as

$$J(f) = \left[\frac{\partial f_i}{\partial x_j} \right]_{1 \le i \le m, 1 \le j \le n}.$$

If $n > m$, and if the rank of the Jacobi matrix attains its maximal value m, f is locally invertible at that point, by the implicit function theorem.

Partial differential equations can be classified by considering the matrix of coefficients of the highest-order differential operators of the equation. For elliptic partial differential equations this matrix is positive definite, which has decisive influence on the set of possible solutions of the equation in question.

The finite element method is an important numerical method to solve partial differential equations, widely applied in simulating complex physical systems. It attempts to approximate the solution to some equation by piecewise linear functions, where the pieces are chosen with respect to a sufficiently fine grid, which in turn can be recast as a matrix equation.

Probability Theory and Statistics

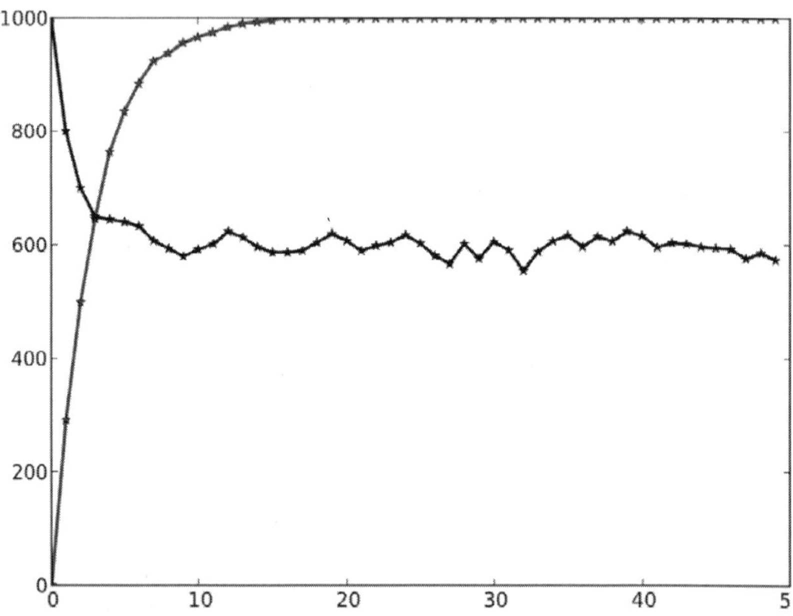

Figure: *Two different Markov chains. The chart depicts the number of particles (of a total of 1000) in state "2". Both limiting values can be determined from the transition matrices, which are given by* $\begin{bmatrix} .7 & 0 \\ .3 & 1 \end{bmatrix}$ *and* $\begin{bmatrix} .7 & .2 \\ .3 & .8 \end{bmatrix}$.

Stochastic matrices are square matrices whose rows are probability vectors, i.e., whose entries are non-negative and sum up to one. Stochastic matrices are used to define Markov chains with finitely many states. A row of the stochastic matrix gives the probability distribution for the next position of some particle currently in the state that corresponds to the row. Properties of the Markov chain like absorbing states, i.e., states that any particle attains eventually, can be read off the eigenvectors of the transition matrices.

Statistics also makes use of matrices in many different forms. Descriptive statistics is concerned with describing data sets, which can often be represented as data matrices, which may then be subjected to dimensionality reduction techniques. The covariance matrix encodes the mutual variance of several random variables. Another technique using matrices are linear least squares, a method that approximates a finite set of pairs (x_1, y_1), (x_2, y_2), ..., (x_N, y_N), by a linear function

$$y_i \lambda ax_i + b, i = 1, ..., N$$

which can be formulated in terms of matrices, related to the singular value decomposition of matrices.

Random matrices are matrices whose entries are random numbers, subject to suitable probability distributions, such as matrix normal distribution. Beyond probability theory, they are applied in domains ranging from number theory to physics.

Symmetries and Transformations in Physics

Linear transformations and the associated symmetries play a key role in modern physics. For example, elementary particles in quantum field theory are classified as representations of the Lorentz group of special relativity and, more specifically, by their behaviour under the spin group. Concrete representations involving the Pauli matrices and more general gamma matrices are an integral part of the physical description of fermions, which behave as spinors. For the three lightest quarks, there is a group-theoretical representation involving the special unitary group SU(3); for their calculations, physicists use a convenient matrix representation known as the Gell-Mann matrices, which are also used for the SU(3) gauge group that forms the basis of the modern description of strong nuclear interactions, quantum chromodynamics. The Cabibbo–Kobayashi–Maskawa matrix, in turn, expresses the fact that the basic quark states that are important for weak interactions are not the same as, but linearly related to the basic quark states that define particles with specific and distinct masses.

Linear Combinations of Quantum States

The first model of quantum mechanics (Heisenberg, 1925) represented the theory's operators by infinite-dimensional matrices acting on quantum states. This is also referred to as matrix mechanics. One particular example is the density matrix that characterizes the "mixed" state of a quantum system as a linear combination of elementary, "pure" eigenstates.

Another matrix serves as a key tool for describing the scattering experiments that form the cornerstone of experimental particle physics: Collision reactions such as occur in particle accelerators, where non-interacting particles head towards each other and collide in a small interaction zone, with a new set of non-interacting particles as the result, can be described as the scalar product of outgoing particle states and a linear combination of ingoing particle states. The linear combination is given by a matrix known as the S-matrix, which encodes all information about the possible interactions between particles.

Normal Modes

A general application of matrices in physics is to the description of linearly coupled harmonic systems. The equations of motion of such systems can be described in matrix form, with a mass matrix multiplying a generalized velocity to give the kinetic term, and a force matrix multiplying a displacement vector to characterize the interactions. The best way to obtain solutions is to determine the system's eigenvectors, its normal modes, by diagonalizing the matrix equation. Techniques like this are crucial when it comes to the internal dynamics of molecules: the internal vibrations of systems consisting of mutually bound component atoms. They are also needed for describing mechanical vibrations, and oscillations in electrical circuits.

Geometrical Optics

Geometrical optics provides further matrix applications. In this approximative theory, the wave nature of light is neglected. The result is a model in which light rays are indeed geometrical rays. If the deflection of light rays by optical elements is small, the action of a lens or reflective element on a given light ray can be expressed as multiplication of a two-component vector with a two-by-two matrix called ray transfer matrix: the vector's components are the light ray's slope and its distance from the optical axis, while the matrix encodes the properties of the optical element. Actually, there are two kinds of matrices, viz. a *refraction matrix* describing the refraction at a lens surface, and a *translation matrix,* describing the translation of the plane of reference to the next refracting surface, where another refraction matrix applies. The optical system, consisting of a combination of lenses and/or reflective elements, is simply described by the matrix resulting from the product of the components' matrices.

Electronics

Traditional mesh analysis in electronics leads to a system of linear equations that can be described with a matrix.

The behaviour of many electronic components can be described using matrices. Let A be a 2-dimensional vector with the component's input voltage v_1 and input current i_1 as its elements, and let B be a 2-dimensional vector with the component's output voltage v_2 and output current i_2 as its elements. Then the behaviour of the electronic component can be described by $B = H \cdot A$, where H is a 2 x 2 matrix containing one impedance element (h_{12}), one admittance element (h_{21}) and two dimensionless elements (h_{11} and h_{22}). Calculating a circuit now reduces to multiplying matrices.

History

Matrices have a long history of application in solving linear equations but they were known as arrays until the 1800s. The Chinese text The Nine Chapters on the Mathematical Art written in 10th–2nd century BCE is the first example of the use of array methods to solve simultaneous equations, including the concept of determinants. In 1545 Italian mathematician Girolamo Cardano brought the method to Europe when he published *Ars Magna*. The Japanese mathematician Seki used the same array methods to solve simultaneous equations in 1683. The Dutch Mathematician Jan de Witt represented transformations using arrays in his 1659 book *Elements of Curves* (1659). Between 1700 and 1710 Gottfried Wilhelm Leibniz publicized the use of arrays for recording information or solutions and experimented with over 50 different systems of arrays. Cramer presented his rule in 1750.

The term "matrix" (Latin for "womb", derived from mater—mother) was coined by James Joseph Sylvester in 1850, who understood a matrix as an object giving rise to a number of determinants today called minors, that is to say, determinants of smaller matrices that derive from the original one by removing columns and rows. In an 1851 paper, Sylvester explains:

> *I have in previous papers defined a "Matrix" as a rectangular array of terms, out of which different systems of determinants may be engendered as from the womb of a common parent.*

Arthur Cayley published a treatise on geometric transformations using matrices that were not rotated versions of the coefficients being investigated as had previously been done. Instead he defined operations such as addition, subtraction, multiplication, and division as transformations of those matrices and showed the associative and distributive properties held true. Cayley investigated and demonstrated the non-commutative property of matrix multiplication as well as the commutative property of matrix addition. Early matrix theory had limited the use of arrays almost exclusively to determinants and Arthur Cayley's abstract matrix operations were revolutionary. He was instrumental in proposing a matrix concept independent of equation systems. In 1858 Cayley published his *Memoir on the theory of matrices* in which he proposed and demonstrated the Cayley-Hamilton theorem.

An English mathematician named Cullis was the first to use modern bracket notation for matrices in 1913 and he simultaneously demonstrated the first significant use the notation $A = [a_{i,j}]$ to represent a matrix where $a_{i,j}$ refers to the ith row and the jth column.

The study of determinants sprang from several sources. Number-theoretical problems led Gauss to relate coefficients of quadratic forms, i.e., expressions such as $x^2 + xy \rightarrow 2y^2$, and linear maps in three dimensions to matrices. Eisenstein further developed these notions, including the remark that, in modern parlance, matrix products are non-commutative. Cauchy was the first to prove general statements about determinants, using as definition of the determinant of a matrix $A = [a_{i,j}]$ the following: replace the powers a_j^k by a_{jk} in the polynomial

$$a_1 a_2 \cdots a_n \prod_{i<j} (a_j - a_i) \text{ ,}$$

where Π denotes the product of the indicated terms. He also showed, in 1829, that the eigenvalues of symmetric matrices are real. Jacobi studied "functional determinants"—later called Jacobi determinants by Sylvester—which can be used to describe geometric transformations at a local (or infinitesimal) level; Kronecker's *Vorlesungen über die Theorie der Determinanten* and Weierstrass' *Zur Determinantentheorie*, both published in 1903, first treated determinants axiomatically, as opposed to previous more concrete approaches such as the mentioned formula of Cauchy. At that point, determinants were firmly established.

Many theorems were first established for small matrices only, for example the Cayley–Hamilton theorem was proved for 2×2 matrices by Cayley in the aforementioned memoir, and by Hamilton for 4×4 matrices. Frobenius, working on bilinear forms, generalized the theorem to all dimensions (1898). Also at the end of the 19th century the Gauss–Jordan elimination (generalizing a special case now known as Gauss elimination) was established by Jordan. In the early 20th century, matrices attained a central role in linear algebra. Partially due to their use in classification of the hypercomplex number systems of the previous century.

The inception of matrix mechanics by Heisenberg, Born and Jordan led to studying matrices with infinitely many rows and columns. Later, von Neumann carried out the mathematical formulation of quantum mechanics, by further developing functional analytic notions such as linear operators on Hilbert spaces, which, very roughly speaking, correspond to Euclidean space, but with an infinity of independent directions.

Other Historical Usages of the Word "Matrix" in Mathematics

The word has been used in unusual ways by at least two authors of historical importance.

Bertrand Russell and Alfred North Whitehead in their *Principia Mathematica* (1910–1913) use the word "matrix" in the context of their Axiom of reducibility. They proposed this axiom as a means to reduce any function to one of lower type, successively, so that at the "bottom" (0 order) the function is identical to its extension:

> *"Let us give the name of* matrix *to any function, of however many variables, which does not involve any apparent variables. Then any possible function other than a matrix is derived from a matrix by means of generalization, i.e., by considering the proposition which asserts that the function in question is true with all possible values or with some value of one of the arguments, the other argument or arguments remaining undetermined".*

For example a function $\vartheta(x, y)$ of two variables x and y can be reduced to a *collection* of functions of a single variable, e.g., y, by "considering" the function for all possible values of "individuals" a_i substituted in place of variable x. And then the resulting collection of functions of the single variable y, i.e., $\forall a_i : \Phi(a_i, y)$, can be reduced to a "matrix" of values by "considering" the function for all possible values of "individuals" b_i substituted in place of variable y: $\forall b_j \forall a_i : \Phi(a_i, b_j)$..

Alfred Tarski in his 1946 *Introduction to Logic* used the word "matrix" synonymously with the notion of truth table as used in mathematical logic.

Chapter 6

Eigenvalues and Eigenvectors

An eigenvector of a square matrix A is a non-zero vector v that, when the matrix multiplies v, yields the same as when some scalar multiplies , the scalar multiplier often being denoted by . That is:

$$Av = \lambda v$$

(Because this equation uses post-multiplication by v, it describes a right eigenvector.)

The number λ is called the eigenvalue of A corresponding to .

If two-dimensional space is visualized as a piece of cloth being stretched by the matrix, the eigenvectors would make up the line along the *direction* the cloth is stretched in and the line of cloth at the center of the stretching, whose direction isn't changed by the stretching either. The eigenvalues for the first line would give the *scale* to which the cloth is stretched, and for the second line the scale to which it is tightened. A reflection may be viewed as stretching a line to scale –1 while shrinking the axis of reflection to scale 1. For 3D rotations, the eigenvectors form the axis of rotation, and since the scale of the axis is unchanged by the rotation, their eigenvalues are all 1.

In analytic geometry, for example, a three-coordinate vector may be seen as an arrow in three-dimensional space starting at the origin. In that case, an eigenvector is an arrow whose direction is either preserved or exactly reversed after multiplication by A. The corresponding eigenvalue determines how the length of the arrow is changed by the operation, and whether its direction is reversed or not, determined by whether the eigenvalue is negative or positive.

In abstract linear algebra, these concepts are naturally extended to more general situations, where the set of real scalar factors $\mathbb{R}^n$ is replaced by any field of scalars (such as algebraic or complex numbers); the set of Cartesian vectors is replaced by any vector space (such as the continuous functions, the polynomials or the trigonometric series), and matrix multiplication is replaced by any linear operator that maps vectors to vectors (such as the derivative from calculus). In such cases, the "vector" in "eigenvector" may be replaced by a more specific term, such as "eigenfunction", "eigenmode", "eigenface", or "eigenstate". Thus, for example, the exponential function $f(x) = e^{\lambda x}$ is an eigenfunction of the derivative operator, , with eigenvalue , since its derivative is .

The set of all eigenvectors of a matrix (or linear operator), each paired with its corresponding eigenvalue, is called the eigensystem of that matrix. Any multiple of an eigenvector is also an eigenvector, with the same eigenvalue. An eigenspace of a matrix A is the set of all eigenvectors with the same eigenvalue, together with the zero vector. An eigenbasis for A is any basis for the set of all vectors that consists of linearly independent eigenvectors of A. Not every matrix has an eigenbasis, but every symmetric matrix does.

The terms characteristic vector, characteristic value, and characteristic space are also used for these concepts. The prefix eigen- is adopted from the German word *eigen* for "own-" or "unique to", "peculiar to", or "belonging to" in the sense of "idiosyncratic" in relation to the originating matrix. Eigenvalues and eigenvectors have many applications in both pure and applied mathematics. They are used in matrix factorization, in quantum mechanics, and in many other areas.

Eigenvectors and Eigenvalues of a Real Matrix

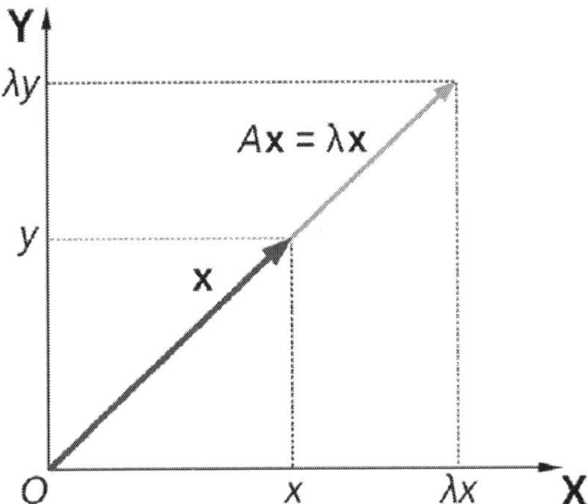

Figure: *Matrix A acts by stretching the vector x, not changing its direction, so x is an eigenvector of A.*

In many contexts, a vector can be assumed to be a list of real numbers (called *coordinates*), written vertically with brackets around the entire list, such as the vectors u and v below. Two vectors are said to be scalar multiples of each other (also called parallel or collinear) if they have the same number of coordinates, and if every coordinate of one vector is obtained by multiplying each corresponding coordinate in the other vector by the same number (known as a *scaling factor*, or a *scalar*). For example, the vectors

$$u = \begin{bmatrix} 1 \\ 3 \\ 4 \end{bmatrix} \text{ and } v = \begin{bmatrix} -20 \\ -60 \\ -80 \end{bmatrix}$$

are scalar multiples of each other, because each coordinate of v is –20 times the corresponding coordinate of u.

A vector with three coordinates, like u or v above, may represent a point in three-dimensional space, relative to some Cartesian coordinate system. It helps to think of such a vector as the tip of an arrow whose tail is at the origin of the coordinate system. In this case, the condition "u is parallel to v" means that the two arrows lie on the same straight line, and may differ only in length and direction along that line.

If we multiply any square matrix A with n rows and n columns by such a vector v, the result will be another vector $w = Av$, also with n rows and one column. That is,

$$\begin{bmatrix} v_1 \\ v_2 \\ \vdots \\ v_n \end{bmatrix} \text{ is mapped to } \begin{bmatrix} w_1 \\ w_2 \\ \vdots \\ w_n \end{bmatrix} = \begin{bmatrix} A_{1,1} & A_{1,2} & \cdots & A_{1,n} \\ A_{2,1} & A_{2,2} & \cdots & A_{2,n} \\ \vdots & \vdots & \ddots & \vdots \\ A_{n,1} & A_{n,2} & \cdots & A_{n,n} \end{bmatrix} \begin{bmatrix} v_1 \\ v_2 \\ \vdots \\ v_n \end{bmatrix}$$

where, for each index ,

$$w_i = A_{i,1}v_1 + A_{i,2}v_2 + \cdots + A_{i,n}v_n = \sum_{j=1}^{n} A_{i,j}v_j$$

In general, if v_j are not all zeros, the vectors and Av will not be parallel. When they *are* parallel (that is, when there is some real number λ such that $Av = \lambda v$) we say that v is an eigenvector of A. In that case, the scale factor λ is said to be the eigenvalue corresponding to that eigenvector.

In particular, multiplication by a 3×3 matrix A may change both the direction and the magnitude of an arrow v in three-dimensional space. However, if is an eigenvector of with eigenvalue , the operation may only change its length, and either keep its direction or flip it (make the arrow point in the exact opposite direction). Specifically, the length of the arrow will increase if , remain the same if , and decrease it if . Moreover, the direction will be precisely the same if , and flipped if . If , then the length of the arrow becomes zero.

An Example

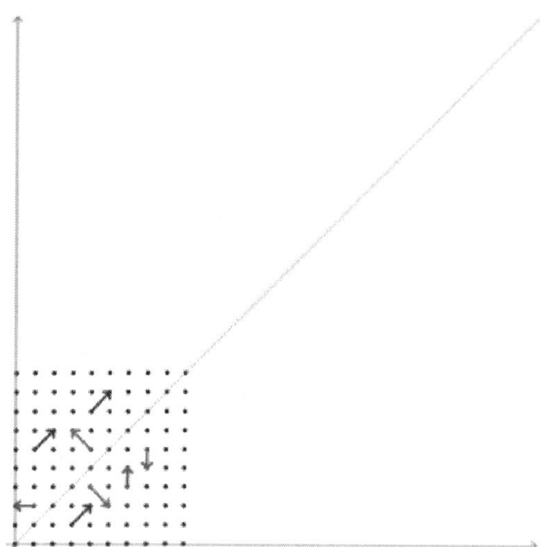

Figure: *The transformation matrix* $\begin{bmatrix} 2 & 1 \\ 1 & 2 \end{bmatrix}$ *preserves the angle of arrows parallel to the lines from the origin to* $\begin{bmatrix} 1 \\ 1 \end{bmatrix}$ *(in blue) and to* $\begin{bmatrix} 1 \\ -1 \end{bmatrix}$ *(in purple). The points that lie on a line through the origin and an eigenvector remain on the line after the transformation. The arrows in red are not parallel to such a line, therefore their angle is altered by the transformation.*

For the transformation matrix

$$A = \begin{bmatrix} 3 & 1 \\ 1 & 3 \end{bmatrix},$$

the vector

$$v = \begin{bmatrix} 4 \\ -4 \end{bmatrix}$$

is an eigenvector with eigenvalue 2. Indeed,

$$Av = \begin{bmatrix} 3 & 1 \\ 1 & 3 \end{bmatrix}\begin{bmatrix} 4 \\ -4 \end{bmatrix} = \begin{bmatrix} 3\cdot4+1\cdot(-4) \\ 1\cdot4+3\cdot(-4) \end{bmatrix} = \begin{bmatrix} 8 \\ -8 \end{bmatrix} = 2\cdot\begin{bmatrix} 4 \\ -4 \end{bmatrix}.$$

On the other hand the vector

$$v = \begin{bmatrix} 0 \\ 1 \end{bmatrix}$$

is *not* an eigenvector, since

$$\begin{bmatrix} 3 & 1 \\ 1 & 3 \end{bmatrix}\begin{bmatrix} 0 \\ 1 \end{bmatrix} = \begin{bmatrix} 3\cdot0+1\cdot1 \\ 1\cdot0+3\cdot1 \end{bmatrix} = \begin{bmatrix} 1 \\ 3 \end{bmatrix},$$

and this vector is not a multiple of the original vector .

Another Example

For the matrix

$$A = \begin{bmatrix} 2 & 0 & 1 \\ 0 & 3 & 0 \\ 1 & 0 & 2 \end{bmatrix},$$

we have

$$A\begin{bmatrix} 1 \\ 0 \\ -1 \end{bmatrix} = \begin{bmatrix} 1 \\ 0 \\ -1 \end{bmatrix} = 1\cdot\begin{bmatrix} 1 \\ 0 \\ -1 \end{bmatrix},$$

$$A\begin{bmatrix} 1 \\ n \\ 1 \end{bmatrix} = \begin{bmatrix} 3 \\ 3n \\ 3 \end{bmatrix} = 3\cdot\begin{bmatrix} 1 \\ n \\ 1 \end{bmatrix}.$$

Therefore, the vectors $[1,0,-1]^T$ and $[1,n,1]^T$ are eigenvectors of A corresponding to the eigenvalues 1 and 3 respectively. (Here the symbol ^{T} indicates matrix transposition, in this case turning the row vectors into column vectors.)

Eigenbasis

An eigenbasis for a linear operator T that operates on a vector space is a basis for that consists entirely of eigenvectors of (possibly with different eigenvalues). Such a basis exists precisely if the direct sum of the eigenspaces equals the whole space, in which case one can take the union of bases chosen in each of the eigenspaces as eigenbasis.

The matrix of T in a given basis is diagonal precisely when that basis is an eigenbasis for T, and for this reason T is called diagonalizable if it admits an eigenbasis.

Generalizations to Infinite-Dimensional Spaces

The definition of eigenvalue of a linear transformation T remains valid even if the underlying space V is an infinite dimensional Hilbert or Banach space. Namely, a scalar λ is an eigenvalue if and only if there is some nonzero vector such that $T(v) = \lambda v$.

Eigenfunctions

A widely used class of linear operators acting on infinite dimensional spaces are the differential operators on function spaces. Let D be a linear differential operator in on the space $\mathbf{C}^{\infty}$ of infinitely differentiable real functions of a real argument t. The eigenvalue equation for D is the differential equation

$$Df = \lambda f$$

The functions that satisfy this equation are commonly called eigenfunctions of D. For the derivative operator d / dt, an eigenfunction is a function that, when differentiated, yields a constant times the original function. The solution is an exponential function

$$f(t) = Ae^{\lambda t},$$

including when λ is zero when it becomes a constant function. Eigenfunctions are an essential tool in the solution of differential equations and many other applied and theoretical fields. For instance, the exponential functions are eigenfunctions of the shift operators. This is the basis of Fourier transform methods for solving problems.

Spectral Theory

If λ is an eigenvalue of T, then the operator $T - \lambda I$ is not one-to-one, and therefore its inverse $(T - \lambda I)^{-1}$ does not exist. The converse is true for finite-dimensional vector spaces, but not for infinite-dimensional ones. In general, the operator $T - \lambda I$ may not have an inverse, even if is not an eigenvalue.

For this reason, in functional analysis one defines the spectrum of a linear operator T as the set of all scalars for which the operator $T - \lambda I$ has no bounded inverse. Thus the spectrum of an operator always contains all its eigenvalues, but is not limited to them.

Left and Right (Algebra)

In algebra, the terms left and right denote the order of a binary operation (usually, but not always called “multiplication”) in non-commutative algebraic structures. A binary operation ” is usually written in the infix form:

$$s * t$$

The argument s is placed on the left side, and the argument t is on the right side. Even if symbol of the operation is omitted, the order of s and t does matter unless “ is commutative.

A two-sided property is fulfilled on both sides. An one-sided property is related to one (unspecified) of two sides.

Although terms are similar, left–right distinction in algebraic parlance is not related neither to left and right limits in calculus, nor to left and right in geometry.

Binary Operation as an Operator

A binary operation ” may be considered as a family of unary operators through currying

$$R_t(s) = s * t,$$

depending on t as a parameter. It is the family of *right* operations. Similarly,

$$L_s(t) = s * t$$

defines the family of *left* operations parametrized with s.

If for some e, the left operation L_e is identical, then e is called a left identity. Similarly, if $R_e = id$, then e is a right identity.

In ring theory, a subring which is invariant under *any* left multiplication in a ring, is called a left ideal. Similarly, a right multiplications-invariant subring is a right ideal.

Left and Right Modules

Over non-commutative rings, the left–right distinction is applied to modules, namely to specify the side where a scalar (module element) appear in the scalar multiplication.

Left module	*Right module*
$s(x + y) = sx + sy$	$(x + y)t = xt + yt$
$(s_1 + s_2)x = s_1x + s_2x$	$x(t_1 + t_2) = xt_1 + xt_2$
$s(tx) = (st)x$	$(xs)t = x(st)$

The distinction is not purely syntactical because implies two different associativity rules (the lowest row in the table) which link multiplication in a module with multiplication in a ring.

A bimodule is simultaneously a left and right module, with two *different* scalar multiplication operations, obeying an obvious associativity condition on them.

Left and Right Eigenvectors

The use of matrices with a single column (rather than a single row) to represent vectors is traditional in many disciplines. For that reason, the word "eigenvector" almost always means a right eigenvector, namely a *column* vector that must be placed to the *right* of the matrix A in the defining equation

$$Av = \lambda v$$

There may be also single-*row* vectors that are unchanged when they occur on the *left* side of a product with a square matrix ; that is, which satisfy the equation

$$uA = \lambda u$$

Any such row vector u is called a left eigenvector of A.

The left eigenvectors of A are transposes of the right eigenvectors of the transposed matrix A^{T}, since their defining equation is equivalent to

$$A^{\mathsf{T}} u^{\mathsf{T}} = \lambda u^{\mathsf{T}}$$

It follows that, if is Hermitian, its left and right eigenvectors are complex conjugates. In particular if is a real symmetric matrix, they are the same except for transposition.

Variational Characterization

In the Hermitian case, eigenvalues can be given a variational characterization. The largest eigenvalue of H is the maximum value of the quadratic form $x^T Hx / x^T x$. A value of that realises that maximum, is an eigenvector.

Computing the Eigenvalues

The eigenvalues of a matrix A can be determined by finding the roots of the characteristic polynomial. Explicit algebraic formulas for the roots of a polynomial exist only if the degree is 4 or less. According to the Abel–Ruffini theorem there is no general, explicit and exact algebraic formula for the roots of a polynomial with degree 5 or more.

It turns out that any polynomial with degree n is the characteristic polynomial of some companion matrix of order n. Therefore, for matrices of order 5 or more, the eigenvalues and eigenvectors cannot be obtained by an explicit algebraic formula, and must therefore be computed by approximate numerical methods.

In theory, the coefficients of the characteristic polynomial can be computed exactly, since they are sums of products of matrix elements; and there are algorithms that can find all the roots of a polynomial of arbitrary degree to any required accuracy. However, this approach is not viable in practice because the coefficients would be contaminated by unavoidable round-off errors, and the roots of a polynomial can be an extremely sensitive function of the coefficients (as exemplified by Wilkinson's polynomial).

Efficient, accurate methods to compute eigenvalues and eigenvectors of arbitrary matrices were not known until the advent of the QR algorithm in 1961. Combining the Householder transformation with the LU decomposition results in an algorithm with better convergence than the QR algorithm.[citation needed] For large Hermitian sparse matrices, the Lanczos algorithm is one example of an efficient iterative method to compute eigenvalues and eigenvectors, among several other possibilities.

Computing the Eigenvectors

Once the (exact) value of an eigenvalue is known, the corresponding eigenvectors can be found by finding non-zero solutions of the eigenvalue equation, that becomes a system of linear equations with known coefficients. For example, once it is known that 6 is an eigenvalue of the matrix

$$A = \begin{bmatrix} 4 & 1 \\ 6 & 3 \end{bmatrix}$$

we can find its eigenvectors by solving the equation $Av = 6v$, that is

$$\begin{bmatrix} 4 & 1 \\ 6 & 3 \end{bmatrix}\begin{bmatrix} x \\ y \end{bmatrix} = 6 \cdot \begin{bmatrix} x \\ y \end{bmatrix}$$

This matrix equation is equivalent to two linear equations

$$\begin{cases} 4x + y &= 6x \\ 6x + 3y &= 6y \end{cases} \text{that is } \begin{cases} -2x + y &= 0 \\ +6x - 3y &= 0 \end{cases}$$

Both equations reduce to the single linear equation $y = 2x$. Therefore, any vector of the form $[a, 2a]'$, for any non-zero real number a, is an eigenvector of A with eigenvalue $\lambda = 6$.

The matrix A above has another eigenvalue $\lambda = 1$. A similar calculation shows that the corresponding eigenvectors are the non-zero solutions of $3x + y = 0$, that is, any vector of the form $[b, -3b]'$, for any non-zero real number b.

Some numeric methods that compute the eigenvalues of a matrix also determine a set of corresponding eigenvectors as a by-product of the computation.

History

Eigenvalues are often introduced in the context of linear algebra or matrix theory. Historically, however, they arose in the study of quadratic forms and differential equations.

In the 18th century Euler studied the rotational motion of a rigid body and discovered the importance of the principal axes. Lagrange realised that the principal axes are the eigenvectors of the inertia matrix. In the early 19th century, Cauchy saw how their work could be used to classify the quadric surfaces, and generalized it to arbitrary dimensions. Cauchy also coined the term *racine caractéristique* (characteristic root) for what is now called *eigenvalue*; his term survives in characteristic equation.

Fourier used the work of Laplace and Lagrange to solve the heat equation by separation of variables in his famous 1822 book Théorie analytique de la chaleur. Sturm developed Fourier's ideas further and brought them to the attention of Cauchy, who combined them with his own ideas and arrived at the fact that real symmetric matrices have real eigenvalues.

This was extended by Hermite in 1855 to what are now called Hermitian matrices. Around the same time, Brioschi proved that the eigenvalues of orthogonal matrices lie on the unit circle, and Clebsch found the corresponding result for skew-symmetric matrices. Finally, Weierstrass clarified an important aspect in the stability theory started by Laplace by realising that defective matrices can cause instability.

In the meantime, Liouville studied eigenvalue problems similar to those of Sturm; the discipline that grew out of their work is now called Sturm–Liouville theory. Schwarz studied the first eigenvalue of Laplace's equation on general domains towards the end of the 19th century, while Poincaré studied Poisson's equation a few years later.

At the start of the 20th century, Hilbert studied the eigenvalues of integral operators by viewing the operators as infinite matrices. He was the first to use the German word *eigen* which means "own", to denote eigenvalues and eigenvectors in 1904, though he may have been following a related usage by Helmholtz. For some time, the standard term in English was "proper value", but the more distinctive term "eigenvalue" is standard today.

The first numerical algorithm for computing eigenvalues and eigenvectors appeared in 1929, when Von Mises published the power method.

One of the most popular methods today, the QR algorithm, was proposed independently by John G.F. Francis and Vera Kublanovskaya in 1961.

Schrodinger Equation

An example of an eigenvalue equation where the transformation $T$is represented in terms of a differential operator is the time-independent Schrφdinger equation in quantum mechanics:

$$H\psi_E = E\psi_E$$

where H, the Hamiltonian, is a second-order differential operator and ψ_E, the wavefunction, is one of its eigenfunctions corresponding to the eigenvalue E, interpreted as its energy.

However, in the case where one is interested only in the bound state solutions of the Schrodinger equation, one looks for ψ_E within the space of square integrable functions. Since this space is a Hilbert space with a well-defined scalar product, one ψ_E can introduce a basis set in which H and can be represented as a one-dimensional array and a matrix respectively. This allows one to represent the Schrödinger equation in a matrix form.

The bra–ket notation is often used in this context. A vector, which represents a state of the system, in the Hilbert space of square integrable functions is represented by $|\Psi_E\rangle$. In this notation, the Schrödinger equation is:

$$H|\Psi_E\rangle = E|\Psi_E\rangle$$

where $|\Psi_E\rangle$ is an eigenstate of H and E represents the eigenvalue. It is a self adjoint operator, the infinite dimensional analog of Hermitian matrices. As in the matrix case, in the equation above $H|\Psi_E\rangle$ is understood to be the vector obtained by application of the transformation H to $|\Psi_E\rangle$.

Molecular Orbitals

In quantum mechanics, and in particular in atomic and molecular physics, within the Hartree–Fock theory, the atomic and molecular orbitals can be defined by the eigenvectors of the Fock operator. The corresponding eigenvalues are interpreted as ionization potentials via Koopmans' theorem. In this case, the term eigenvector is used in a somewhat more general meaning, since the Fock operator is explicitly dependent on the orbitals and their eigenvalues. If one wants to underline this aspect one speaks of nonlinear eigenvalue problem. Such equations are usually solved by an iteration procedure, called in this case self-consistent field method. In quantum chemistry, one often represents the Hartree–Fock equation in a non-orthogonal basis set. This particular representation is a generalized eigenvalue problem called Roothaan equations.

Geology and Glaciology

In geology, especially in the study of glacial till, eigenvectors and eigenvalues are used as a method by which a mass of information of a clast fabric's constituents' orientation and dip can be summarized in a 3-D space by six numbers. In the field, a geologist may collect such data for hundreds or thousands of clasts in a soil sample, which can only be compared graphically such as in a Tri-Plot (Sneed and Folk) diagram, or as a Stereonet on a Wulff Net.

The output for the orientation tensor is in the three orthogonal (perpendicular) axes of space. The three eigenvectors are ordered v_1, v_2, v_3 by their eigenvalues $E_1 \geq E_2 \geq E_3$; then is the primary orientation/dip of clast, v_2 is the secondary and v_3 is the tertiary, in terms of strength. The clast orientation is defined as the direction of the eigenvector, on a compass rose of 360°. Dip is measured as the eigenvalue, the modulus of the tensor: this is valued from 0° (no dip) to 90° (vertical). The relative values of E_1, E_2, E_3 and are dictated by the nature of the sediment's fabric. If $E_1 = E_2 = E_3$, the fabric is said to be isotropic. If $E_1 = E_2 > E_3$, the fabric is said to be planar. If $E_1 > E_2 > E_3$, the fabric is said to be linear.

Principal Components Analysis

The eigendecomposition of a symmetric positive semidefinite (PSD) matrix yields an orthogonal basis of eigenvectors, each of which has a nonnegative eigenvalue. The orthogonal decomposition of a PSD matrix is used in multivariate analysis, where the sample covariance matrices are PSD. This orthogonal decomposition is called principal components analysis (PCA) in statistics.

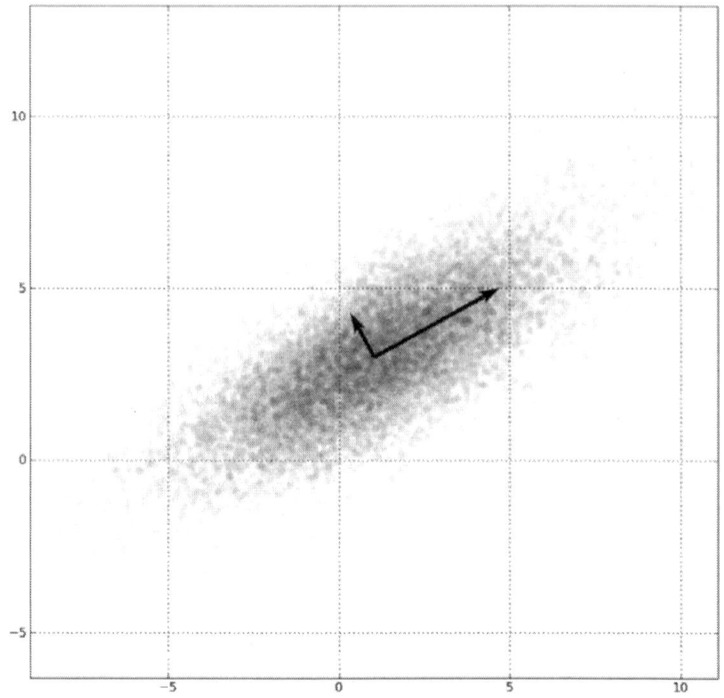

Figure: *PCA of the multivariate Gaussian distribution centered at* $(1,3)$ *with a standard deviation of 3 in roughly the* $(0.878, 0.478)$ *direction and of 1 in the orthogonal direction. The vectors shown are unit eigenvectors of the (symmetric, positive-semidefinite) covariance matrix scaled by the square root of the corresponding eigenvalue. (Just as in the one-dimensional case, the square root is taken because the standard deviation is more readily visualized than the variance.*

PCA studies linear relations among variables. PCA is performed on the covariance matrix or the correlation matrix (in which each variable is scaled to have its sample variance equal to one). For the covariance or correlation matrix, the eigenvectors correspond to principal components and the eigenvalues to the variance explained by the principal components. Principal component analysis of the correlation matrix provides an orthonormal eigen-basis for the space of the observed data: In this basis, the largest eigenvalues correspond to the principal-components that are associated with most of the covariability among a number of observed data.

Principal component analysis is used to study large data sets, such as those encountered in data mining, chemical research, psychology, and in marketing. PCA is popular especially in psychology, in the field of psychometrics.

In Q methodology, the eigenvalues of the correlation matrix determine the Q-methodologist's judgment of *practical* significance (which differs from the statistical significance of hypothesis testing; cf. criteria for determining the number of factors). More generally, principal component analysis can be used as a method of factor analysis in structural equation modelling.

Vibration Analysis

Eigenvalue problems occur naturally in the vibration analysis of mechanical structures with many degrees of freedom. The eigenvalues are the natural frequencies (or eigenfrequencies) of vibration, and the eigenvectors are the shapes of these vibrational modes. In particular, undamped vibration is governed by

$$m\ddot{x} + kx = 0$$

or

$$m\ddot{x} = -kx$$

that is, acceleration is proportional to position (i.e., we expect x to be sinusoidal in time).

In n dimensions, m becomes a mass matrix and k a stiffness matrix. Admissible solutions are then a linear combination of solutions to the generalized eigenvalue problem

$$-kx = \omega^2 mx$$

where ω^2 is the eigenvalue and ω is the angular frequency. Note that the principal vibration modes are different from the principal compliance modes, which are the eigenvectors of k alone. Furthermore, damped vibration, governed by

$$m\ddot{x} + c\dot{x} + kx = 0$$

leads to what is called a so-called quadratic eigenvalue problem,

$$(\omega^2 m + \omega c + k)x = 0.$$

This can be reduced to a generalized eigenvalue problem by clever use of algebra at the cost of solving a larger system.

The orthogonality properties of the eigenvectors allows decoupling of the differential equations so that the system can be represented as linear summation of the eigenvectors. The eigenvalue problem of complex structures is often solved using finite element analysis, but neatly generalize the solution to scalar-valued vibration problems.

Eigenfaces

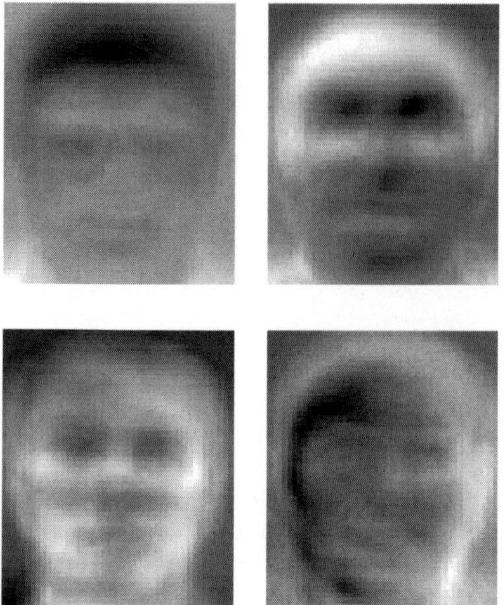

Figure: *Eigenfaces* as examples of eigenvectors

In image processing, processed images of faces can be seen as vectors whose components are the brightnesses of each pixel.[26] The dimension of this vector space is the number of pixels. The eigenvectors of the covariance matrix associated with a large set of normalized pictures of faces are called eigenfaces; this is an example of principal components analysis. They are very useful for expressing any face image as a linear combination of some of them. In the facial recognition branch of biometrics, eigenfaces provide a means of applying data compression to faces for identification purposes. Research related to eigen vision systems determining hand gestures has also been made.

Similar to this concept, eigenvoices represent the general direction of variability in human pronunciations of a particular utterance, such as a word in a language. Based on a linear combination of such eigenvoices, a new voice pronunciation of the word can be constructed. These concepts have been found useful in automatic speech recognition systems, for speaker adaptation.

Tensor of Moment of Inertia

In mechanics, the eigenvectors of the moment of inertia tensor define the principal axes of a rigid body. The tensor of moment of inertia is a key quantity required to determine the rotation of a rigid body around its center of mass.

Stress Tensor

In solid mechanics, the stress tensor is symmetric and so can be decomposed into a diagonal tensor with the eigenvalues on the diagonal and eigenvectors as a basis. Because it is diagonal, in this orientation, the stress tensor has no shear components; the components it does have are the principal components.

Eigenvalues of a Graph

In spectral graph theory, an eigenvalue of a graph is defined as an eigenvalue of the graph's adjacency matrix A, or (increasingly) of the graph's Laplacian matrix, which is either $T - A$ (sometimes called the *combinatorial Laplacian*) or $I - T^{-1/2}AT^{-1/2}$ (sometimes called the *normalized Laplacian*), where T is a diagonal matrix with T_{ii} equal to the degree of vertex v_i, and in $T^{-1/2}$, the ith diagonal entry is $1/\sqrt{\deg(v_i)}$. The kth principal eigenvector of a graph is defined as either the eigenvector corresponding to the th largest or th smallest eigenvalue of the Laplacian. The first principal eigenvector of the graph is also referred to merely as the principal eigenvector.

The principal eigenvector is used to measure the centrality of its vertices. An example is Google's PageRank algorithm. The principal eigenvector of a modified adjacency matrix of the World Wide Web graph gives the page ranks as its components. This vector corresponds to the stationary distribution of the Markov chain represented by the row-normalized adjacency matrix; however, the adjacency matrix must first be modified to ensure a stationary distribution exists. The second smallest eigenvector can be used to partition the graph into clusters, via spectral clustering. Other methods are also available for clustering.

Basic Reproduction Number

The basic reproduction number (R_0) is a fundamental number in the study of how infectious diseases spread. If one infectious person is put into a population of completely susceptible people, then R_0 is the average number of people that one typical infectious person will infect.

The generation time of an infection is the time, t_G, from one person becoming infected to the next person becoming infected. In a heterogeneous population, the next generation matrix defines how many people in the population will become infected after time t_G has passed. t_G is then the largest eigenvalue of the next generation matrix.

Laplace Transform

The Laplace transform is a widely used integral transform in mathematics with many applications in physics and engineering. It is a linear operator of a function $f(t)$ with a real argument t ($t \geq 0$) that transforms $f(t)$ to a function $F(s)$ with complex argument s, given by the integral

$$F(s) = \int_0^\infty f(t)e^{-st}\,dt.$$

This transformation is bijective for the majority of practical uses; the most-common pairs of $f(t)$ and $F(s)$ are often given in tables for easy reference. The Laplace transform is named after Pierre-Simon Laplace, who introduced it in his work on probability theory.

The Laplace transform is related to the Fourier transform, but whereas the Fourier transform expresses a function or signal as a superposition of sinusoids, the Laplace transform expresses a function, more generally, as a superposition of moments. Like the Fourier transform, the Laplace transform is used for solving differential and integral equations. In physics and engineering it is used for analysis of linear time-invariant systems such as electrical circuits, harmonic oscillators, optical devices, and mechanical systems. In such analyses, the Laplace transform is often interpreted as a transformation from the time-domain, in which inputs and outputs are functions of time, to the frequency-domain, where the same inputs and outputs are functions of complex angular frequency, in radians per unit time.

Given a simple mathematical or functional description of an input or output to a system, the Laplace transform provides an alternative functional description that often simplifies the process of analyzing the behaviour of the system, or in synthesizing a new system based on a set of specifications. So, for example, Laplace transformation from the time domain to the frequency domain transforms differential equations into algebraic equations and convolution into multiplication.

History

The Laplace transform is named after mathematician and astronomer Pierre-Simon Laplace, who used a similar transform (now called z transform) in his work on probability theory. The current widespread use of the transform came about soon after World War II although it had been used in the 19th century by Abel, Lerch, Heaviside, and Bromwich. The older history of similar transforms is as follows. From 1744, Leonhard Euler investigated integrals of the form

$$z = \int X(x) e^{ax} dx \quad \text{and} \quad z = \int X(x) x^{A} dx$$

as solutions of differential equations but did not pursue the matter very far. Joseph Louis Lagrange was an admirer of Euler and, in his work on integrating probability density functions, investigated expressions of the form

$$\int X(x) e^{-ax} a^{x} dx,$$

which some modern historians have interpreted within modern Laplace transform theory.

These types of integrals seem first to have attracted Laplace's attention in 1782 where he was following in the spirit of Euler in using the integrals themselves as solutions of equations. However, in 1785, Laplace took the critical step forward when, rather than just looking for a solution in the form of an integral, he started to apply the transforms in the sense that was later to become popular. He used an integral of the form:

$$\int x^{s} \phi(x) dx,$$

akin to a Mellin transform, to transform the whole of a difference equation, in order to look for solutions of the transformed equation. He then went on to apply the Laplace transform in the same way and started to derive some of its properties, beginning to appreciate its potential power.

Laplace also recognised that Joseph Fourier's method of Fourier series for solving the diffusion equation could only apply to a limited region of space as the solutions were periodic. In 1809, Laplace applied his transform to find solutions that diffused indefinitely in space.

Formal Definition

The Laplace transform of a function $f(t)$, defined for all real numbers $t \geq 0$, is the function $F(s)$, defined by:

$$F(s) = \int_0^{\infty} e^{-st} f(t) dt$$

The parameter s is a complex number:

$s = \sigma + i\omega$, with real numbers σ and ω.

Other notations for the Laplace transform include $\mathcal{L}f$ or alternatively $\mathcal{L}_t\{f(t)\}$ instead of F.

The meaning of the integral depends on types of functions of interest. A necessary condition for existence of the integral is that f

must be locally integrable on $[0, \infty)$. For locally integrable functions that decay at infinity or are of exponential type, the integral can be understood as a (proper) Lebesgue integral. However, for many applications it is necessary to regard it as a conditionally convergent improper integral at ∞. Still more generally, the integral can be understood in a weak sense, and this is dealt with below.

One can define the Laplace transform of a finite Borel measure μ by the Lebesgue integral

$$(\mathcal{L}\mu)(s) = \int_{[0,\infty)} e^{-st} d\mu(t).$$

An important special case is where μ is a probability measure or, even more specifically, the Dirac delta function. In operational calculus, the Laplace transform of a measure is often treated as though the measure came from a distribution function *f*. In that case, to avoid potential confusion, one often writes

$$(\mathcal{L}f)(s) = \int_{0^-}^{\infty} e^{-st} f(t)dt$$

where the lower limit of 0" is shorthand notation for

$$\lim_{\varepsilon \downarrow 0} \int_{-\varepsilon}^{\infty}.$$

This limit emphasizes that any point mass located at 0 is entirely captured by the Laplace transform. Although with the Lebesgue integral, it is not necessary to take such a limit, it does appear more naturally in connection with the Laplace–Stieltjes transform.

Probability Theory

In pure and applied probability, the Laplace transform is defined as an expected value. If *X* is a random variable with probability density function *f*, then the Laplace transform of *f* is given by the expectation

$$(\mathcal{L}f)(s) = E\left[e^{-sX}\right]$$

By abuse of language, this is referred to as the Laplace transform of the random variable *X* itself. Replacing *s* by $-t$ gives the moment generating function of *X*. The Laplace transform has applications throughout probability theory, including first passage times of stochastic processes such as Markov chains, and renewal theory.

Of particular use is the ability to recover the cumulative distribution function of a continuous random variable *X* by means of the Laplace transform as follows

$$F_X(x)=\mathcal{L}_s^{-1}\left\{\frac{E\left[e^{-sX}\right]}{s}\right\}(x)=\mathcal{L}_s^{-1}\left\{\frac{(\mathcal{L}f)(s)}{s}\right\}(x)$$

Two-Sided Laplace Transform

In mathematics, the two-sided Laplace transform or bilateral Laplace transform is an integral transform equivalent to probability's moment generating function. Two-sided Laplace transforms are closely related to the Fourier transform, the Mellin transform, and the ordinary or one-sided Laplace transform. If $f(t)$ is a real or complex valued function of the real variable t defined for all real numbers, then the two-sided Laplace transform is defined by the integral

$$\mathcal{B}\{f(t)\}=F(s)=\int_{-\infty}^{\infty}e^{-st}f(t)dt.$$

The integral is most commonly understood as an improper integral, which converges if and only if each of the integrals

$$\int_0^{\infty}e^{-st}f(t)dt,\quad \int_{-\infty}^{0}e^{-st}f(t)dt$$

exists. There seems to be no generally accepted notation for the two-sided transform; the $\mathcal{B}$ used here recalls "bilateral". The two-sided transform used by some authors is

$$\mathcal{T}\{f(t)\}=s\mathcal{B}\{f\}=sF(s)=s\int_{-\infty}^{\infty}e^{-st}f(t)dt.$$

In pure mathematics the argument t can be any variable, and Laplace transforms are used to study how differential operators transform the function.

In science and engineering applications, the argument t often represents time (in seconds), and the function $f(t)$ often represents a signal or waveform that varies with time. In these cases, the signals are transformed by filters, that work like a mathematical operator, but with a restriction. They have to be causal, which means that the output in a given time t cannot depend on an output which is a higher value of t. In population ecology, the argument t often represents spatial displacement in a dispersal kernel.

When working with functions of time, $f(t)$ is called the time domain representation of the signal, while $F(s)$ is called the s-domain (or *Laplace domain*) representation. The inverse transformation then represents a *synthesis* of the signal as the sum of its frequency components taken over all frequencies, whereas the forward transformation represents the *analysis* of the signal into its frequency components.

Relationship to Other Integral Transforms

If $u(t)$ is the Heaviside step function, equal to zero when t is less than zero, to one-half when t equals zero, and to one when t is greater than zero, then the Laplace transform $\mathcal{L}$ may be defined in terms of the two-sided Laplace transform by

$$\mathcal{L}\{f(t)\} = \mathcal{B}\{f(t)u(t)\}.$$

On the other hand, we also have

$$\{\mathcal{B}f\}(s) = \{\mathcal{L}f(t)\}(s) + \{\mathcal{L}f(-t)\}(-s)$$

so either version of the Laplace transform can be defined in terms of the other.

The Mellin transform may be defined in terms of the two-sided Laplace transform by

$$\{\mathcal{M}f\}(s) = \{\mathcal{B}f(e^{-x})\}(s)$$

and conversely we can get the two-sided transform from the Mellin transform by

$$\{\mathcal{B}f\}(s) = \{\mathcal{M}f(-\ln x)\}(s).$$

The Fourier transform may also be defined in terms of the two-sided Laplace transform; here instead of having the same image with differing originals, we have the same original but different images. We may define the Fourier transform as

$$\mathcal{F}\{f(t)\} = F(s = i\omega) = F(\omega).$$

Note that definitions of the Fourier transform differ, and in particular

$$\{\mathcal{F}f\} = F(s = i\omega) = \frac{1}{\sqrt{2\pi}}\{\mathcal{B}f\}(s)$$

is often used instead. In terms of the Fourier transform, we may also obtain the two-sided Laplace transform, as

$$\{\mathcal{B}f\}(s) = \{\mathcal{F}f\}(-is).$$

The Fourier transform is normally defined so that it exists for real values; the above definition defines the image in a strip $a < \Im(s) < b$ which may not include the real axis.

The moment-generating function of a continuous probability density function $f(x)$ can be expressed as $\{\mathcal{B}f\}(-s)$.

Properties

It has basically the same properties of the unilateral transform with an important difference

Properties of the Unilateral Laplace Transform			
	Time domain	***unilateral-'s' domain***	***bilateral-'s' domain***
Differentiation	$f'(t)$	$sF(s)-f(0)$	$sF(s)$
Second Differentiation	$f''(t)$	$s^2F(s)-sf(0)-f'(0)$	$s^2F(s)$

To use the bilateral transform is equivalent to assume null initial conditions. Therefore it is more suitable than the unilateral for calculating transfer functions from the differential equations, or when looking for an easy particular solution.

Inverse Laplace Transform

Two integrable functions have the same Laplace transform only if they differ on a set of Lebesgue measure zero. This means that, on the range of the transform, there is an inverse transform. In fact, besides integrable functions, the Laplace transform is a one-to-one mapping from one function space into another in many other function spaces as well, although there is usually no easy characterization of the range. Typical function spaces in which this is true include the spaces of bounded continuous functions, the space $L^{\infty}(0,\infty)$, or more generally tempered functions (that is, functions of at worst polynomial growth) on $(0,\infty)$. The Laplace transform is also defined and injective for suitable spaces of tempered distributions.

In these cases, the image of the Laplace transform lives in a space of analytic functions in the region of convergence. The inverse Laplace transform is given by the following complex integral, which is known by various names (the Bromwich integral, the Fourier-Mellin integral, and Mellin's inverse formula):

$$f(t)=\mathcal{L}^{-1}\{F\}=\mathcal{L}_s^{-1}\{F(s)\}\equiv\frac{1}{2\pi i}\lim_{T\to\infty}\int_{\gamma-iT}^{\gamma+iT}e^{st}F(s)ds,$$

where γ is a real number so that the contour path of integration is in the region of convergence of $F(s)$. An alternative formula for the inverse

Laplace transform is given by Post's inversion formula. The limit here is interpreted in the weak-* topology.

In practice it is typically more convenient to decompose a Laplace transform into known transforms of functions obtained from a table, and construct the inverse by inspection.

Region of Convergence

If f is a locally integrable function (or more generally a Borel measure locally of bounded variation), then the Laplace transform $F(s)$ of f converges provided that the limit

$$\lim_{R\to\infty}\int_0^R f(t)e^{-st}\,dt$$

exists. The Laplace transform converges absolutely if the integral

$$\int_0^\infty \left|f(t)e^{-st}\right|dt$$

exists (as a proper Lebesgue integral). The Laplace transform is usually understood as conditionally convergent, meaning that it converges in the former instead of the latter sense.

The set of values for which $F(s)$ converges absolutely is either of the form $\mathrm{Re}(s) > a$ or else $\mathrm{Re}(s) \geq a$, where a is an extended real constant, $-\infty \leq a \leq \infty$. (This follows from the dominated convergence theorem.) The constant a is known as the abscissa of absolute convergence, and depends on the growth behaviour of $f(t)$. Analogously, the two-sided transform converges absolutely in a strip of the form $a < \mathrm{Re}(s) < b$, and possibly including the lines $\mathrm{Re}(s) = a$ or $\mathrm{Re}(s) = b$. The subset of values of s for which the Laplace transform converges absolutely is called the region of absolute convergence or the domain of absolute convergence. In the two-sided case, it is sometimes called the strip of absolute convergence. The Laplace transform is analytic in the region of absolute convergence.

Similarly, the set of values for which $F(s)$ converges (conditionally or absolutely) is known as the region of conditional convergence, or simply the region of convergence (ROC). If the Laplace transform converges (conditionally) at $s = s_0$, then it automatically converges for all s with $\mathrm{Re}(s) > \mathrm{Re}(s_0)$. Therefore the region of convergence is a half-plane of the form $\mathrm{Re}(s) > a$, possibly including some points of the boundary line $\mathrm{Re}(s) = a$. In the region of convergence $\mathrm{Re}(s) > \mathrm{Re}(s_0)$, the Laplace transform of f can be expressed by integrating by parts as the integral

$$F(s) = (s - s_0)\int_0^\infty e^{-(s-s_0)t}\beta(t)dt, \quad \beta(u) = \int_0^u e^{-s_0 t} f(t)dt.$$

That is, in the region of convergence $F(s)$ can effectively be expressed as the absolutely convergent Laplace transform of some other function. In particular, it is analytic.

There are several Paley–Wiener theorems concerning the relationship between the decay properties of f and the properties of the Laplace transform within the region of convergence.

In engineering applications, a function corresponding to a linear time-invariant (LTI) system is *stable* if every bounded input produces a bounded output. This is equivalent to the absolute convergence of the Laplace transform of the impulse response function in the region $\mathrm{Re}(s) \geq 0$. As a result, LTI systems are stable provided the poles of the Laplace transform of the impulse response function have negative real part.

Properties and Theorems

The Laplace transform has a number of properties that make it useful for analyzing linear dynamical systems. The most significant advantage is that differentiation and integration become multiplication and division, respectively, by s (similarly to logarithms changing multiplication of numbers to addition of their logarithms).

Because of this property, the Laplace variable s is also known as *operator variable* in the L domain: either *derivative operator* or (for s^{-1}) *integration operator*. The transform turns integral equations and differential equations to polynomial equations, which are much easier to solve. Once solved, use of the inverse Laplace transform reverts to the time domain.

- Initial value theorem: $f(0^+) = \lim_{s\to\infty} sF(s)$.
- Final value theorem: $f(\infty) = \lim_{s\to 0} sF(s)$, if all poles of $sF(s)$ are in the left half-plane.

The final value theorem is useful because it gives the long-term behaviour without having to perform partial fraction decompositions or other difficult algebra. If a function has poles in the right-hand plane or on the imaginary axis, (e.g. e^t or $\sin(t)$, respectively) the behaviour of this formula is undefined.

Relation to Power Series

The Laplace transform can be viewed as a continuous analogue of a power series. If $a(n)$ is a discrete function of a positive integer n, then the power series associated to $a(n)$ is the series

$$\sum_{n=0}^{\infty} a(n)x^n$$

where x is a real variable. Replacing summation over n with integration over t, a continuous version of the power series becomes

$$\int_0^{\infty} f(t)x^t \, dt$$

where the discrete function $a(n)$ is replaced by the continuous one $f(t)$. Changing the base of the power from x to e gives

$$\int_0^{\infty} f(t)\left(e^{\log x}\right)^t dt$$

For this to converge for, say, all bounded functions f, it is necessary to require that $\log x < 0$. Making the substitution $-s = \log x$ gives just the Laplace transform:

$$\int_0^{\infty} f(t)e^{-st} \, dt$$

In other words, the Laplace transform is a continuous analog of a power series in which the discrete parameter n is replaced by the continuous parameter t, and x is replaced by e^{-s}.

Relation to Moments

The quantities

$$\mu_n = \int_0^{\infty} t^n f(t)dt$$

are the *moments* of the function f. If the first n moments of f converge absolutely, then by repeated differentiation under the integral, $(-1)^n (\mathcal{L}f)^{(n)}(0) = \mu_n$.

This is of special significance in probability theory, where the moments of a random variable X are given by the expectation values $\mu_n = E[X^n]$. Then the relation holds:

$$\mu_n = (-1)^n \frac{d^n}{ds^n} E\left[e^{-sX}\right].$$

Proof of the Laplace Transform of a Function's Derivative

It is often convenient to use the differentiation property of the Laplace transform to find the transform of a function's derivative. This can be derived from the basic expression for a Laplace transform as follows:

$$\begin{aligned}\mathcal{L}\{f(t)\} &= \int_{0^-}^{\infty} e^{-st} f(t)dt \\ &= \left[\frac{f(t)e^{-st}}{-s}\right]_{0^-}^{\infty} - \int_{0^-}^{\infty} \frac{e^{-st}}{-s} f'(t)dt \quad \text{(by parts)} \\ &= \left[-\frac{f(0^-)}{-s}\right] + \frac{1}{s}\mathcal{L}\{f'(t)\},\end{aligned}$$

yielding

$$\mathcal{L}\{f'(t)\} = s\cdot\mathcal{L}\{f(t)\} - f(0^-),$$

and in the bilateral case,

$$\mathcal{L}\{f'(t)\} = s\int_{-\infty}^{\infty} e^{-st} f(t)dt = s\cdot\mathcal{L}\{f(t)\}.$$

The general result

$$\mathcal{L}\left\{f^{(n)}(t)\right\} = s^n\cdot\mathcal{L}\{f(t)\} - s^{n-1}f(0^-) - \cdots - f^{(n-1)}(0^-),$$

where $f^{(n)}$ denotes the nth derivative of f, can then be established with an inductive argument.

Evaluating Improper Integrals

Let $\mathcal{L}\{f(t)\} = F(s)$, then

$$\mathcal{L}\left\{\frac{f(t)}{t}\right\} = \int_s^{\infty} F(p)dp,$$

or

$$\int_0^{\infty} \frac{f(t)}{t} e^{-st}\, dt = \int_s^{\infty} F(p)dp.$$

Letting $s \to 0$, gives one the identity

$$\int_0^{\infty} \frac{f(t)}{t} dt = \int_0^{\infty} F(p)dp.$$

provided that the interchange of limits can be justified. Even when the interchange cannot be justified the calculation can be suggestive. For example, proceeding formally one has

$$\int_0^\infty \frac{\cos at - \cos bt}{t} dt = \int_0^\infty \left(\frac{p}{p^2 + a^2} - \frac{p}{p^2 + b^2} \right) dp = \frac{1}{2} \ln \frac{p^2 + a^2}{p^2 + b^2} \Big|_0^\infty = \ln b - \ln a.$$

The validity of this identity can be proved by other means. It is an example of a Frullani integral.

Another example is Dirichlet integral.

Relationship to Other Transforms

Laplace–Stieltjes Transform: The (unilateral) Laplace–Stieltjes transform of a function $g : \mathrm{R} \to \mathrm{R}$ is defined by the Lebesgue–Stieltjes integral

$$\{\mathcal{L}^* g\}(s) = \int_0^\infty e^{-st} dg(t).$$

The function g is assumed to be of bounded variation. If g is the antiderivative of f:

$$g(x) = \int_0^x f(t) dt$$

then the Laplace–Stieltjes transform of g and the Laplace transform of f coincide. In general, the Laplace–Stieltjes transform is the Laplace transform of the Stieltjes measure associated to g.

So in practice, the only distinction between the two transforms is that the Laplace transform is thought of as operating on the density function of the measure, whereas the Laplace–Stieltjes transform is thought of as operating on its cumulative distribution function.

Fourier Transform

The continuous Fourier transform is equivalent to evaluating the bilateral Laplace transform with imaginary argument $s = i\omega$ or $s = 2\partial fi$:

$$\begin{aligned} \hat{f}(\omega) &= \mathcal{F}\{f(t)\} \\ &= \mathcal{L}\{f(t)\}|_{s=i\omega} = F(s)|_{s=i\omega} \\ &= \int_{-\infty}^\infty e^{-i\omega t} f(t) dt. \end{aligned}$$

This definition of the Fourier transform requires a prefactor of 1/ 2π on the reverse Fourier transform. This relationship between the Laplace and Fourier transforms is often used to determine the frequency spectrum of a signal or dynamical system.

The above relation is valid as stated if and only if the region of convergence (ROC) of $F(s)$ contains the imaginary axis, $\sigma = 0$. For example, the function $f(t) = \cos(\omega_0 t)$ has a Laplace transform $F(s) = s/(s^2 + \omega_0^2)$ whose ROC is $\mathrm{Re}(s) > 0$. As $s = i\omega$ is a pole of $F(s)$, substituting $s = i\omega$ in $F(s)$ does not yield the Fourier transform of $f(t)u(t)$, which is proportional to the Dirac delta-function $\delta(\omega\text{-}\omega_0)$.

However, a relation of the form

$$\lim_{\sigma \to 0^+} F(\sigma + i\omega) = \hat{f}(\omega)$$

holds under much weaker conditions. For instance, this holds for the above example provided that the limit is understood as a weak limit of measures. General conditions relating the limit of the Laplace transform of a function on the boundary to the Fourier transform take the form of Paley-Wiener theorems.

Mellin Transform

The Mellin transform and its inverse are related to the two-sided Laplace transform by a simple change of variables. If in the Mellin transform

$$G(s) = \mathcal{M}\{g(\theta)\} = \int_0^\infty \theta^s g(\theta) \frac{d\theta}{\theta}$$

we set $\theta = e^{-t}$ we get a two-sided Laplace transform.

Z-Transform

The unilateral or one-sided Z-transform is simply the Laplace transform of an ideally sampled signal with the substitution of

$$z \overset{\text{def}}{=} e^{sT}$$

where $T = 1/f_s$ is the sampling period (in units of time e.g., seconds) and f_s is the sampling rate (in samples per second or hertz)

Let

$$\Delta_T(t) \overset{\text{def}}{=} \sum_{n=0}^{\infty} \delta(t - nT)$$

be a sampling impulse train (also called a Dirac comb) and

$$\begin{aligned} x_q(t) &\overset{\text{def}}{=} x(t)\Delta_T(t) = x(t)\sum_{n=0}^{\infty} \delta(t - nT) \\ &= \sum_{n=0}^{\infty} x(nT)\delta(t - nT) = \sum_{n=0}^{\infty} x[n]\delta(t - nT) \end{aligned}$$

be the sampled representation of the continuous-time $x(t)$

$$x[n] \stackrel{\text{def}}{=} x(nT)$$

The Laplace transform of the sampled signal $x_q(t)$ is

$$\begin{aligned} X_q(s) &= \int_{0^-}^{\infty} x_q(t)e^{-st}\,dt \\ &= \int_{0^-}^{\infty} \sum_{n=0}^{\infty} x[n]\delta(t-nT)e^{-st}\,dt \\ &= \sum_{n=0}^{\infty} x[n] \int_{0^-}^{\infty} \delta(t-nT)e^{-st}\,dt \\ &= \sum_{n=0}^{\infty} x[n]e^{-nsT}. \end{aligned}$$

This is the precise definition of the unilateral Z-transform of the discrete function $x[n]$

$$X(z) = \sum_{n=0}^{\infty} x[n]z^{-n}$$

with the substitution of $z \to e^{sT}$.

Comparing the last two equations, we find the relationship between the unilateral Z-transform and the Laplace transform of the sampled signal:

$$X_q(s) = X(z)\Big|_{z=e^{sT}}.$$

The similarity between the Z and Laplace transforms is expanded upon in the theory of time scale calculus.

Borel Transform

The integral form of the Borel transform

$$F(s) = \int_0^{\infty} f(z)e^{-sz}\,dz$$

is a special case of the Laplace transform for f an entire function of exponential type, meaning that

$$|f(z)| \leq Ae^{B|z|}$$

for some constants A and B. The generalized Borel transform allows a different weighting function to be used, rather than the exponential function, to transform functions not of exponential type. Nachbin's theorem gives necessary and sufficient conditions for the Borel transform to be well defined.

Fundamental Relationships

Since an ordinary Laplace transform can be written as a special case of a two-sided transform, and since the two-sided transform can be written as the sum of two one-sided transforms, the theory of the Laplace-, Fourier-, Mellin-, and Z-transforms are at bottom the same subject. However, a different point of view and different characteristic problems are associated with each of these four major integral transforms.

s-Domain Equivalent Circuits and Impedances

The Laplace transform is often used in circuit analysis, and simple conversions to the s-Domain of circuit elements can be made. Circuit elements can be transformed into impedances, very similar to phasor impedances.

Here is a summary of equivalents:

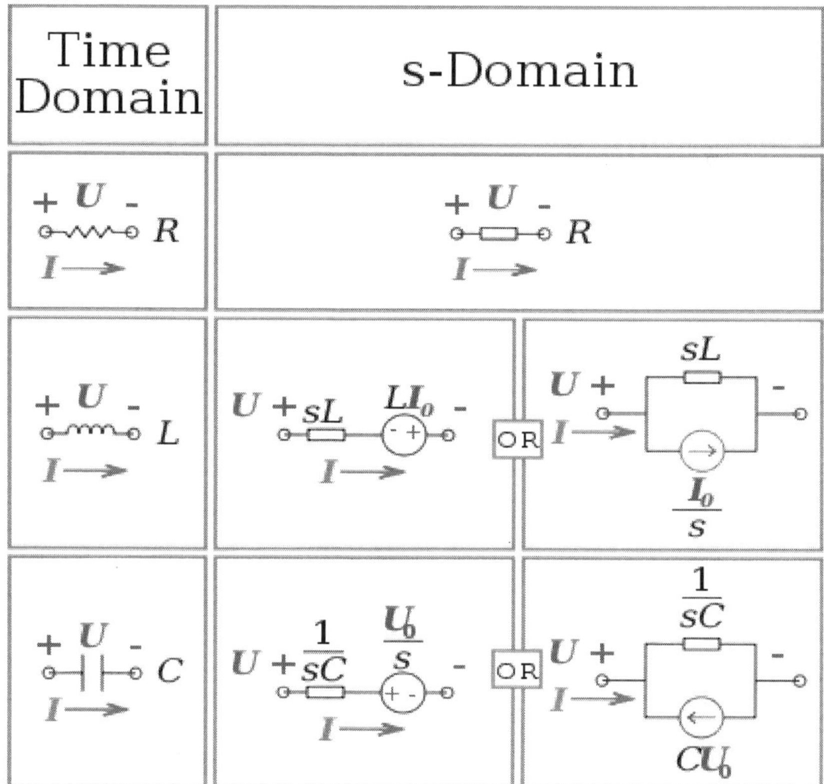

Note that the resistor is exactly the same in the time domain and the s-Domain. The sources are put in if there are initial conditions on the circuit elements.

For example, if a capacitor has an initial voltage across it, or if the inductor has an initial current through it, the sources inserted in the s-Domain account for that.

The equivalents for current and voltage sources are simply derived from the transformations in the table above.

Examples: How to Apply the Properties and Theorems

The Laplace transform is used frequently in engineering and physics; the output of a linear time invariant system can be calculated by convolving its unit impulse response with the input signal. Performing this calculation in Laplace space turns the convolution into a multiplication; the latter being easier to solve because of its algebraic form.

The Laplace transform can also be used to solve differential equations and is used extensively in electrical engineering. The Laplace transform reduces a linear differential equation to an algebraic equation, which can then be solved by the formal rules of algebra. The original differential equation can then be solved by applying the inverse Laplace transform.

The English electrical engineer Oliver Heaviside first proposed a similar scheme, although without using the Laplace transform; and the resulting operational calculus is credited as the Heaviside calculus.

Example 1: Solving a Differential Equation

In nuclear physics, the following fundamental relationship governs radioactive decay: the number of radioactive atoms N in a sample of a radioactive isotope decays at a rate proportional to N. This leads to the first order linear differential equation

$$\frac{dN}{dt} = -\lambda N$$

where λ is the decay constant. The Laplace transform can be used to solve this equation.

Rearranging the equation to one side, we have

$$\frac{dN}{dt} + \lambda N = 0.$$

Next, we take the Laplace transform of both sides of the equation:

$$\left(s\tilde{N}(s) - N_o\right) + \lambda\tilde{N}(s) = 0$$

where

$$\tilde{N}(s) = \mathcal{L}\{N(t)\}$$

and

$$N_o = N(0).$$

Solving, we find

$$\tilde{N}(s) = \frac{N_o}{s+\lambda}.$$

Finally, we take the inverse Laplace transform to find the general solution

$$\begin{aligned} N(t) &= \mathcal{L}^{-1}\{\tilde{N}(s)\} = \mathcal{L}^{-1}\left\{\frac{N_o}{s+\lambda}\right\} \\ &= N_o e^{-\lambda t}, \end{aligned}$$

which is indeed the correct form for radioactive decay.

Example: Deriving the Complex Impedance for a Capacitor

In the theory of electrical circuits, the current flow in a capacitor is proportional to the capacitance and rate of change in the electrical potential (in SI units). Symbolically, this is expressed by the differential equation

$$i = C\frac{dv}{dt}$$

where C is the capacitance (in farads) of the capacitor, $i = i(t)$ is the electric current (in amperes) through the capacitor as a function of time, and $v = v(t)$ is the voltage (in volts) across the terminals of the capacitor, also as a function of time.

Taking the Laplace transform of this equation, we obtain

$$I(s) = C\left(sV(s) - V_o\right)$$

where

$$\begin{aligned} I(s) &= \mathcal{L}\{i(t)\}, \\ V(s) &= \mathcal{L}\{v(t)\}, \end{aligned}$$

and

$$V_o = v(t)\big|_{t=0} .$$

Solving for $V(s)$ we have

$$V(s) = \frac{I(s)}{sC} + \frac{V_o}{s}.$$

The definition of the complex impedance Z (in ohms) is the ratio of the complex voltage V divided by the complex current I while holding the initial state V_o at zero:

$$Z(s) = \frac{V(s)}{I(s)}\bigg|_{V_o=0} .$$

Using this definition and the previous equation, we find:

$$Z(s) = \frac{1}{sC},$$

which is the correct expression for the complex impedance of a capacitor.

Example: Method of Partial Fraction Expansion

Consider a linear time-invariant system with transfer function

$$H(s) = \frac{1}{(s+\alpha)(s+\beta)}.$$

The impulse response is simply the inverse Laplace transform of this transfer function:

$$h(t) = \mathcal{L}^{-1}\{H(s)\}.$$

To evaluate this inverse transform, we begin by expanding $H(s)$ using the method of partial fraction expansion:

$$\frac{1}{(s+\alpha)(s+\beta)} = \frac{P}{s+\alpha} + \frac{R}{s+\beta}.$$

The unknown constants P and R are the residues located at the corresponding poles of the transfer function. Each residue represents the relative contribution of that singularity to the transfer function's overall shape.

By the residue theorem, the inverse Laplace transform depends only upon the poles and their residues. To find the residue P, we multiply both sides of the equation by $s + \alpha$ to get

$$\frac{1}{s+\beta} = P + \frac{R(s+\alpha)}{s+\beta}.$$

Then by letting $s = -\alpha$, the contribution from R vanishes and all that is left is

$$P = \left.\frac{1}{s+\beta}\right|_{s=-\alpha} = \frac{1}{\beta-\alpha}.$$

Similarly, the residue R is given by

$$R = \left.\frac{1}{s+\alpha}\right|_{s=-\beta} = \frac{1}{\alpha-\beta}.$$

Note that

$$R = \frac{-1}{\beta-\alpha} = -P$$

and so the substitution of R and P into the expanded expression for $H(s)$ gives

$$H(s) = \left(\frac{1}{\beta-\alpha}\right)\cdot\left(\frac{1}{s+\alpha} - \frac{1}{s+\beta}\right).$$

Finally, using the linearity property and the known transform for exponential decay, we can take the inverse Laplace transform of $H(s)$ to obtain:

$$h(t) = \mathcal{L}^{-1}\{H(s)\} = \frac{1}{\beta-\alpha}\left(e^{-\alpha t} - e^{-\beta t}\right),$$

which is the impulse response of the system.

Example : Convolution

The same result can be achieved using the convolution property as if the system is a series of filters with transfer functions of $1/(s + a)$ and $1/(s + b)$. That is, the inverse of

$$H(s) = \frac{1}{(s+a)(s+b)} = \frac{1}{s+a}\cdot\frac{1}{s+b}$$

is

$$\mathcal{L}^{-1}\left\{\frac{1}{s+a}\right\}*\mathcal{L}^{-1}\left\{\frac{1}{s+b}\right\} = e^{-at}*e^{-bt} = \int_0^t e^{-ax}e^{-b(t-x)}\,dx = \frac{e^{-at}-e^{-bt}}{b-a}.$$

Example: Mixing Sines, Cosines, and Exponentials

Time function

$$e^{-\alpha t}\left[\cos(\omega t)+\left(\frac{\beta-\alpha}{\omega}\right)\sin(\omega t)\right]u(t)$$

Laplace transform

$$\frac{s+\beta}{(s+\alpha)^2+\omega^2}$$

Starting with the Laplace transform

$$X(s)=\frac{s+\beta}{(s+\alpha)^2+\omega^2},$$

we find the inverse transform by first adding and subtracting the same constant α to the numerator:

$$X(s)=\frac{s+\alpha}{(s+\alpha)^2+\omega^2}+\frac{\beta-\alpha}{(s+\alpha)^2+\omega^2}.$$

By the shift-in-frequency property, we have

$$\begin{aligned} x(t) &= e^{-\alpha t}\mathcal{L}^{-1}\left\{\frac{s}{s^2+\omega^2}+\frac{\beta-\alpha}{s^2+\omega^2}\right\} \\ &= e^{-\alpha t}\mathcal{L}^{-1}\left\{\frac{s}{s^2+\omega^2}+\left(\frac{\beta-\alpha}{\omega}\right)\left(\frac{\omega}{s^2+\omega^2}\right)\right\} \\ &= e^{-\alpha t}\left[\mathcal{L}^{-1}\left\{\frac{s}{s^2+\omega^2}\right\}+\left(\frac{\beta-\alpha}{\omega}\right)\mathcal{L}^{-1}\left\{\frac{\omega}{s^2+\omega^2}\right\}\right]. \end{aligned}$$

Finally, using the Laplace transforms for sine and cosine, we have

$$x(t) = e^{-\alpha t}\left[\cos(\omega t)u(t)+\left(\frac{\beta-\alpha}{\omega}\right)\sin(\omega t)u(t)\right].$$

$$x(t) = e^{-\alpha t}\left[\cos(\omega t)+\left(\frac{\beta-\alpha}{\omega}\right)\sin(\omega t)\right]u(t).$$

Example : Phase Delay

Time function

$$\sin(\omega t+\phi)$$

Laplace transform

$$\frac{s\sin\phi + \omega\cos\phi}{s^2 + \omega^2}$$

$$\frac{s\cos\phi - \omega\sin\phi}{s^2 + \omega^2}$$

Starting with the Laplace transform,

$$X(s) = \frac{s\sin\phi + \omega\cos\phi}{s^2 + \omega^2}$$

we find the inverse by first rearranging terms in the fraction:

$$\begin{aligned} X(s) &= \frac{s\sin\phi}{s^2 + \omega^2} + \frac{\omega\cos\phi}{s^2 + \omega^2} \\ &= (\sin\phi)\left(\frac{s}{s^2 + \omega^2}\right) + (\cos\phi)\left(\frac{\omega}{s^2 + \omega^2}\right). \end{aligned}$$

We are now able to take the inverse Laplace transform of our terms:

$$\begin{aligned} x(t) &= (\sin\phi)\mathcal{L}^{-1}\left\{\frac{s}{s^2 + \omega^2}\right\} + (\cos\phi)\mathcal{L}^{-1}\left\{\frac{\omega}{s^2 + \omega^2}\right\} \\ &= (\sin\phi)(\cos\omega t) + (\sin\omega t)(\cos\phi). \end{aligned}$$

This is just the sine of the sum of the arguments, yielding:

$$x(t) = \sin(\omega t + \phi).$$

We can apply similar logic to find that

$$\mathcal{L}^{-1}\left\{\frac{s\cos\phi - \omega\sin\phi}{s^2 + \omega^2}\right\} = \cos(\omega t + \phi).$$

Example: Determining Structure of Astronomical Object from Spectrum

The wide and general applicability of the Laplace transform and its inverse is illustrated by an application in astronomy which provides some information on the *spatial distribution* of matter of an astronomical source of radiofrequency thermal radiation too distant to resolve as more than a point, given its flux density spectrum, rather than relating the *time* domain with the spectrum (frequency domain).

Assuming certain properties of the object, e.g. spherical shape and constant temperature, calculations based on carrying out an inverse Laplace transformation on the spectrum of the object can produce the only possible model of the distribution of matter in it (density as a function of distance from the center) consistent with the spectrum. When independent information on the structure of an object is available, the inverse Laplace transform method has been found to be in good agreement.

Chapter 7

Line and Plane Equations

Equation of a Line

An important topic of high school algebra is "the equation of a line." This means an equation in x and y whose solution set is a line in the (x, y) plane.

The most popular form in algebra is the "slope-intercept" form

$$y = mx + b.$$

This in effect uses x as a parameter and writes y as a function of x: $y = f(x) = mx + b$. When $x = 0$, $y = b$ and the point (0,b) is the intersection of the line with the y-axis.

Thinking of a line as a geometrical object and not the graph of a function, it makes sense to treat x and y more evenhandedly. The general equation for a line (normal form) is

$$ax + by = c,$$

with the stipulation that at least one of a or b is nonzero. This can easily be converted to slope-intercept form by solving for y:

$$y = (-a/b)x + c/b,$$

except for the special case $b = 0$, when the line is parallel to the y-axis.

If the coefficients on the normal form are multiplied by a nonzero constant, the set of solutions is exactly the same, so, for example, all these equations have the same line as solution.

$$2x + 3y = 4$$

$$4x + 6y = 8$$

$$-x - (3/2)y = -2$$

$(1/2)x + (3/4)y = 1$

In general, if k is a nonzero constant, then these are equations for the same line, since they have the same solutions.

$ax + by = c$

$(ka)x + (kb)y = kc.$

A popular choice for k, in the case when c is not zero, is $k = (1/c)$. Then the equation becomes

$(a/c)x + (b/c)y = 1.$

Another useful form of the equation is to divide by $|(a,b)|$, the square root of $a^2 + b^2$.

Finding the Equation of a Line through 2 Points in the Plane

For any two points P and Q, there is exactly one line PQ through the points. If the coordinates of P and Q are known, then the coefficients a, b, c of an equation for the line can be found by solving a system of linear equations.

Example: For P = (1, 2), Q = (-2, 5), find the equation ax + by = c of line PQ.

Since P is on the line, its coordinates satisfy the equation: a1 + b2 = c, or a + 2b = c.

Since Q is on the line, its coordinates satisfy the equation: a(-2) + b5 = c, or -2 a + 5b = c.

Multiply the first equation by 2 and add to eliminate a from the equation: 4b + 5b = 9b = 2c + c = 3c, so b = (1/3)c. Then substituting into the first equation, a = c - 2b = c - (2/3)c = (1/3)c.

This gives the equation [(1/3)c]x + [(1/3)c}y = c. Why is the c not solved for? Remember that there are an infinite number of equations for the line, each of which is multiple of the other. We can factor out c (or set c = 1 for the same result) and get (1/3)x + (1/3)y =1 as one choice of equation for the line. Another choice might be c = 3: x+y = 3, which has cleared the denominators.

This method always works for any distinct P and Q. There is of course a formula for a, b, c also. This can be found expressed by determinants, or the cross product.

Connection with Parametric Form of a Line

Given two points P and Q, the points of line PQ can be written as F(t) = (1-t)P + tQ, for t ranging over all the real numbers. If both P and Q satisfy the same equation ax+by = c, then a computation shows that this is also true for (1-t)P + tQ, for any choice of t.

Here is this computation. Let $P = (p_1, p_2)$, $Q = (q_1, q_2)$. Then since the points are on the line, we know that both

$$ap_1 + bp_2 = c$$

$$aq_1 + bq_2 = c.$$

For the point F(t), we must check $a[(1-t)p_1+tq_1] + b[(1-t)p_2+tq_2] = c$. But the left side can be rearranged as $(1-t)(ap_1 + bp_2) + t(aq_1 + bq_2)$, and this equals $(1-t)c + tc = c$. So the equation holds. Compare this explicit computation with the computation given for the plane that uses dot product. The computations are the same, but one shows more detail and one hides the coordinates and shows a more conceptual picture.

Equation of a Plane

A plane in 3-space has the equation

$$ax + by + cz = d,$$

where at least one of the numbers a, b, c must be nonzero.

As for the line, if the equation is multiplied by any nonzero constant k to get the equation $kax + kby + kcz = kd$, the plane of solutions is the same. If c is not zero, it is often useful to think of the plane as the graph of a function z of x and y. The equation can be rearranged like this:

$$z = -(a/c)x = (-b/c)\ y + d/c$$

Another useful choice, when d is not zero, is to divide by d so that the constant term = 1.

$$(a/d)x + (b/d)y + (c/d)z = 1.$$

Another useful form of the equation is to divide by $|(a,b,c)|$, the square root of $a^2 + b^2 + c^2$.

Finding the Equation of a Plane through 3 Points in Space

Given points P, Q, R in space, find the equation of the plane through the 3 points.

Example: P = (1, 1, 1), Q = (1, 2, 0), R = (-1, 2, 1). We seek the coefficients of an equation $ax + by + cz = d$, where P, Q and R satisfy the equations, thus:

$$a + b + c = d$$

$$a + 2b + 0c = d$$

$$-a + 2b + c = d$$

Subtracting the first equation from the second and then adding the first equation to the third, we eliminate a to get

$$b - c = 0$$

$$4b + c = 2d$$

Adding the equations gives 5b = 2d, or b = (2/5)d, then solving for c = b = (2/5)d and then a = d - b - c = (1/5)d.

So the equation (with a nonzero constant left in to choose) is d(1/5)x + d(2/5)y + d(2/5)z = d, so one choice of constant gives

$$x + 2y + 2z = 5$$

or another choice would be (1/5)x + (2/5)y + (2/5)z = 1

Given the coordinates of P, Q, R, there is a formula for the coefficients of the plane that uses determinants or cross product.

Vector Fields in Plane and in Space

A vector field in the plane, for instance, can be visualized as a collection of arrows with a given magnitude and direction each attached to a point in the plane. Vector fields are often used to model, for example, the speed and direction of a moving fluid throughout space, or the strength and direction of some force, such as the magnetic or gravitational force, as it changes from point to point.

The elements of differential and integral calculus extend to vector fields in a natural way. When a vector field represents force, the line integral of a vector field represents the work done by a force moving along a path, and under this interpretation conservation of energy is exhibited as a special case of the fundamental theorem of calculus. Vector fields can usefully be thought of as representing the velocity of a moving flow in space, and this physical intuition leads to notions such as the divergence (which represents the rate of change of volume of a flow) and curl (which represents the rotation of a flow).

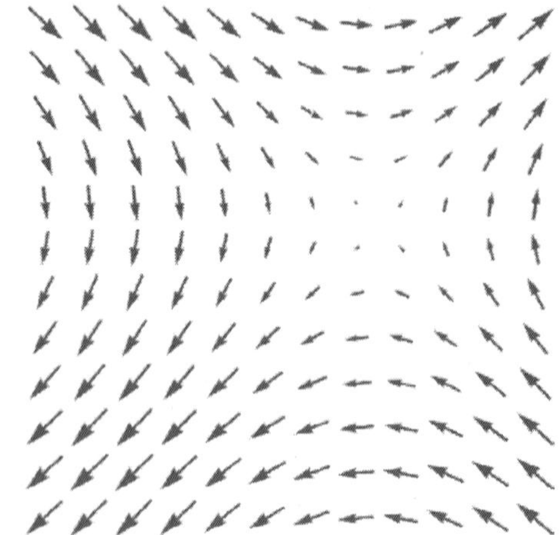

Figure: *A portion of the vector field (sin* y*, sin* x*)*

In coordinates, a vector field on a domain in *n*-dimensional Euclidean space can be represented as a vector-valued function that associates an n-tuple of real numbers to each point of the domain. This representation of a vector field depends on the coordinate system, and there is a well-defined transformation law in passing from one coordinate system to the other. Vector fields are often discussed on open subsets of Euclidean space, but also make sense on other subsets such as surfaces, where they associate an arrow tangent to the surface at each point (a tangent vector).

More generally, vector fields are defined on differentiable manifolds, which are spaces that look like Euclidean space on small scales, but may have more complicated structure on larger scales. In this setting, a vector field gives a tangent vector at each point of the manifold (that is, a section of the tangent bundle to the manifold). Vector fields are one kind of tensor field.

Vector Fields on Subsets of Euclidean Space

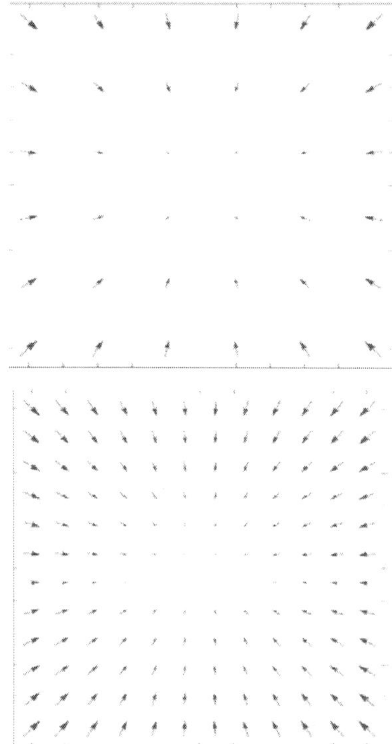

Figure: *Two representations of the same vector field: v(x, y) = –r. The arrows depict the field at discrete points, however, the field exists everywhere.*

Given a subset S in R^n, a vector field is represented by a vector-valued function $V: S \rightarrow \mathrm{R}^n$ in standard Cartesian coordinates (x_1, ..., x_n). If each component of V is continuous, then V is a continuous vector field, and more generally V is a C^k vector field if each component V is k times continuously differentiable.

A vector field can be visualized as assigning a vector to individual points within an n-dimensional space.

Given two C^k-vector fields V, W defined on S and a real valued C^k-function f defined on S, the two operations scalar multiplication and vector addition

$$(fV)(p) := f(p)V(p)$$

$$(V+W)(p) := V(p)+W(p)$$

define the module of C^k-vector fields over the ring of C^k-functions.

Coordinate Transformation Law

In physics, a vector is additionally distinguished by how its coordinates change when one measures the same vector with respect to a different background coordinate system. The transformation properties of vectors distinguish a vector as a geometrically distinct entity from a simple list of scalars, or from a covector.

Thus, suppose that $(x_1,...,x_n)$ is a choice of Cartesian coordinates, in terms of which the components of the vector V are

$$V_x = (V_{1,x}, \ldots, V_{n,x})$$

and suppose that $(y_1,...,y_n)$ are n functions of the x_i defining a different coordinate system. Then the components of the vector V in the new coordinates are required to satisfy the transformation law

$$V_{i,y} = \sum_{j=1}^{n} \frac{\partial y_j}{\partial x_i} V_{j,x}. \tag{1}$$

Such a transformation law is called contravariant. A similar transformation law characterizes vector fields in physics: specifically, a vector field is a specification of n functions in each coordinate system subject to the transformation law (1) relating the different coordinate systems.

Vector fields are thus contrasted with scalar fields, which associate a number or *scalar* to every point in space, and are also contrasted with simple lists of scalar fields, which do not transform under coordinate changes.

Vector Fields on Manifolds

Given a differentiable manifold M, a vector field on M is an assignment of a tangent vector to each point in M. More precisely, a vector field F is a mapping from M into the tangent bundle TM so that $p^\circ F$ is the identity mapping where p denotes the projection from TM to M. In other words, a vector field is a section of the tangent bundle.

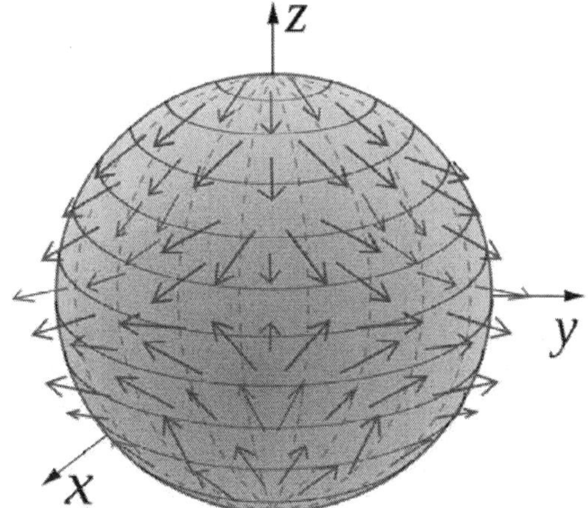

Figure: *A vector field on a sphere*

If the manifold M is smooth or analytic—that is, the change of coordinates is smooth (analytic)—then one can make sense of the notion of smooth (analytic) vector fields. The collection of all smooth vector fields on a smooth manifold M is often denoted by $\Gamma(TM)$ or $C^\infty(M,TM)$ (especially when thinking of vector fields as sections); the collection of all smooth vector fields is also denoted by $\mathfrak{X}(M)$ (a fraktur "X").

Examples

- A vector field for the movement of air on Earth will associate for every point on the surface of the Earth a vector with the wind speed and direction for that point. This can be drawn using arrows to represent the wind; the length (magnitude) of the arrow will be an indication of the wind speed. A "high" on the usual barometric pressure map would then act as a source (arrows pointing away), and a "low" would be a sink (arrows pointing towards), since air tends to move from high pressure areas to low pressure areas.
- Velocity field of a moving fluid. In this case, a velocity vector is associated to each point in the fluid.
- Streamlines, Streaklines and Pathlines are 3 types of lines that can be made from vector fields. They are:

  Streaklines — as revealed in wind tunnels using smoke.

  Streamlines (or fieldlines)— as a line depicting the instantaneous field at a given time.

Pathlines — showing the path that a given particle (of zero mass) would follow.

- Magnetic fields. The fieldlines can be revealed using small iron filings.
- Maxwell's equations allow us to use a given set of initial conditions to deduce, for every point in Euclidean space, a magnitude and direction for the force experienced by a charged test particle at that point; the resulting vector field is the electromagnetic field.
- A gravitational field generated by any massive object is also a vector field. For example, the gravitational field vectors for a spherically symmetric body would all point towards the sphere's center with the magnitude of the vectors reducing as radial distance from the body increases.

Gradient Field

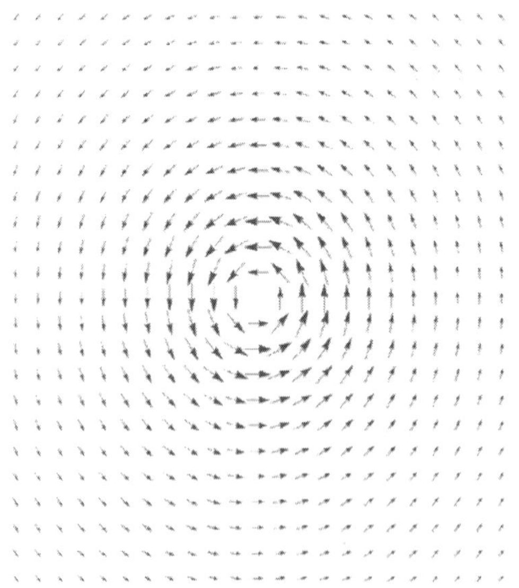

Figure: *A vector field that has circulation about a point cannot be written as the gradient of a function.*

Vector fields can be constructed out of scalar fields using the gradient operator (denoted by the del: ∇).

A vector field V defined on a set S is called a gradient field or a conservative field if there exists a real-valued function (a scalar field) f on S such that

$$V = \nabla f = \left(\frac{\partial f}{\partial x_1}, \frac{\partial f}{\partial x_2}, \frac{\partial f}{\partial x_3}, \ldots, \frac{\partial f}{\partial x_n}\right).$$

The associated flow is called the gradient flow, and is used in the method of gradient descent. The path integral along any closed curve γ (γ(0) = γ(1)) in a gradient field is zero:

$$\int_\gamma \langle V(x), \mathrm{d}x\rangle = \int_\gamma \langle \nabla f(x), \mathrm{d}x\rangle = f(\gamma(1)) - f(\gamma(0)).$$

where the angular brackets and comma: $\langle , \rangle$ denotes the inner product of two vectors (strictly speaking - the integrand $V(x)$ is a 1-form rather than a vector in the elementary sense).

Central Field

A C^∞-vector field over $\mathrm{R}^n \setminus \{0\}$ is called a central field if

$$V(T(p)) = T(V(p)) \qquad (T \in \mathrm{O}(n, \mathbf{R}))$$

where $\mathrm{O}(n, \mathrm{R})$ is the orthogonal group. We say central fields are invariant under orthogonal transformations around 0.

The point 0 is called the center of the field.

Since orthogonal transformations are actually rotations and reflections, the invariance conditions mean that vectors of a central field are always directed towards, or away from, 0; this is an alternate (and simpler) definition. A central field is always a gradient field, since defining it on one semiaxis and integrating gives an antigradient.

Operations on Vector Fields

Line Integral: A common technique in physics is to integrate a vector field along a curve, i.e. to determine its line integral. Given a particle in a gravitational vector field, where each vector represents the force acting on the particle at a given point in space, the line integral is the work done on the particle when it travels along a certain path.

The line integral is constructed analogously to the Riemann integral and it exists if the curve is rectifiable (has finite length) and the vector field is continuous.

Given a vector field V and a curve γ parametrized by $[a, b]$ (where a and b are real) the line integral is defined as

$$\int_\gamma \langle V(x), \mathrm{d}x\rangle = \int_a^b \langle V(\gamma(t)), \gamma'(t)\, \mathrm{d}t\rangle.$$

Divergence

The divergence of a vector field on Euclidean space is a function (or scalar field). In three-dimensions, the divergence is defined by

$$\operatorname{div} \mathbf{F} = \nabla \cdot \mathbf{F} = \frac{\partial F_1}{\partial x} + \frac{\partial F_2}{\partial y} + \frac{\partial F_3}{\partial z},$$

with the obvious generalization to arbitrary dimensions. The divergence at a point represents the degree to which a small volume around the point is a source or a sink for the vector flow, a result which is made precise by the divergence theorem.

The divergence can also be defined on a Riemannian manifold, that is, a manifold with a Riemannian metric that measures the length of vectors.

Curl

The curl is an operation which takes a vector field and produces another vector field. The curl is defined only in three-dimensions, but some properties of the curl can be captured in higher dimensions with the exterior derivative. In three-dimensions, it is defined by

$$\text{curl}\mathbf{F} = \nabla \times \mathbf{F} = \left(\frac{\partial F_3}{\partial y} - \frac{\partial F_2}{\partial z}\right)\mathbf{e}_1 - \left(\frac{\partial F_3}{\partial x} - \frac{\partial F_1}{\partial z}\right)\mathbf{e}_2 + \left(\frac{\partial F_2}{\partial x} - \frac{\partial F_1}{\partial y}\right)\mathbf{e}_3.$$

The curl measures the density of the angular momentum of the vector flow at a point, that is, the amount to which the flow circulates around a fixed axis. This intuitive description is made precise by Stokes' theorem.

Index of a Vector Field

The index of a vector field is a way of describing the behaviour of a vector field around an isolated zero (i.e. non-singular point) which can distinguish saddles from sources and sinks. Take a small sphere around the zero so that no other zeros are included.

A map from this sphere to a unit sphere of dimensions $n-1$ can be constructed by dividing each vector by its length to form a unit length vector which can then be mapped to the unit sphere. The index of the vector field at the point is the degree of this map. The index of the vector field is the sum of the indices of each zero.

The index will be zero around any non singular point, it is +1 around sources and sinks and -1 around saddles. In two dimensions the index is equivalent to the winding number.

For an ordinary sphere in three dimension space it can be shown that the index of any vector field on the sphere must be two, this leads to the hairy ball theorem which shows that every such vector field must have a zero. This theorem generalises to the Poincaré–Hopf theorem which relates the index to the Euler characteristic of the space.

History

Vector fields arose originally in classical field theory in 19th century physics, specifically in magnetism. They were formalized by Michael Faraday, in his concept of lines of force, who emphasized that the field *itself* should be an object of study, which it has become throughout physics in the form of field theory.

In addition to the magnetic field, other phenomena that were modelled as vector fields by Faraday include the electrical field and light field.

Line–Plane Intersection

In analytic geometry, the intersection of a line and a plane can be the empty set, a point, or a line. Distinguishing these cases, and determining equations for the point and line in the latter cases have use, for example, in computer graphics, motion planning, and collision detection.

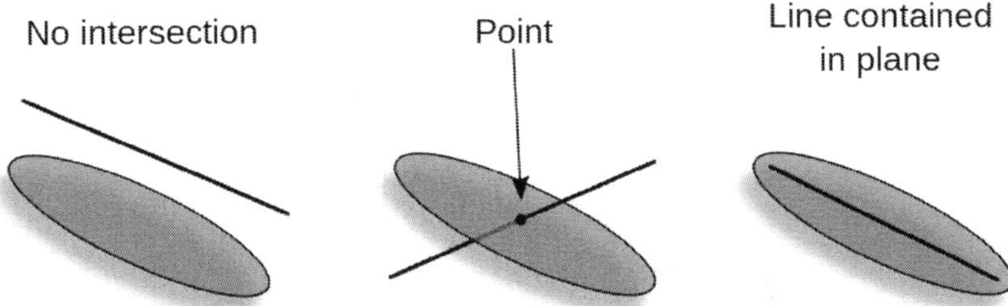

The three possible plane-line intersections:

1. No intersection.
2. Point intersection.
3. Line intersection.

Algebraic Form

In vector notation, a plane can be expressed as the set of points $\mathbf{p}$ for which

$$(\mathbf{p} - \mathbf{p}_0)\cdot\mathbf{n} = 0$$

where n is a normal vector to the plane and P0 is a point on the plane. (The notation a.b denotes the dot product of the two vector a and b.) The vector equation for a line is

$$\mathbf{p} = d\mathbf{l} + \mathbf{l}_0 \quad d \in \mathbb{R}$$

where I is a vector in the direction of the line, $\mathbf{l}_0$ is a point on the line, and d is a scalar in the real number domain. Substitute the equation for the line line into the equation for the plane gives

$$(d\mathbf{l} + \mathbf{l}_0 - \mathbf{p}_0)\cdot\mathbf{n} = 0$$

Expanding gives

$$d\mathbf{l}\cdot\mathbf{n} + (\mathbf{l}_0 - \mathbf{p}_0)\cdot\mathbf{n} = 0$$

And solve for

$$d = \frac{(\mathbf{p}_0 - \mathbf{l}_0)\cdot\mathbf{n}}{\mathbf{l}\cdot\mathbf{n}}.$$

If $\mathbf{l}\cdot\mathbf{n} = 0$ then the line and plane are parallel. There will be two cases if then the line is contained inside the plane, the line intersects the plane everywhere. Otherwise the line and plane have no intersection.

If $\mathbf{l}\cdot\mathbf{n} \neq 0$ there is a single point of intersection. The value of d can be calculated and the point of intersection is given by

Parametric Form

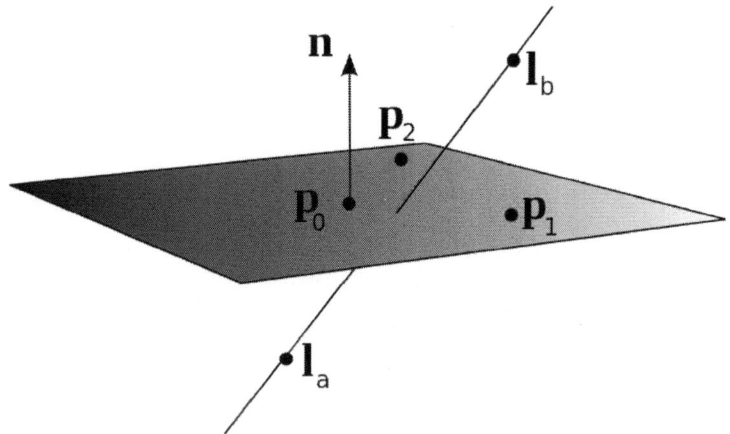

Figure: *The intersection of line and plane.*

A line is described by all points that are a given direction from a point. Thus a general point on a line can be represented as

$$\mathbf{l}_a + (\mathbf{l}_b - \mathbf{l}_a)t, \quad t \in \mathbb{R}$$

where $\mathbf{l}_a = (x_a, y_a, z_a)$ and $\mathbf{l}_b = (x_b, y_b, z_b)$ are two distinct points along the line.

Similarly a general point on a plane can be represented as

$$\mathbf{p}_0 + (\mathbf{p}_1 - \mathbf{p}_0)u + (\mathbf{p}_2 - \mathbf{p}_0)v, \quad u, v \in \mathbb{R}$$

where $\mathbf{p}_k = (x_k, y_k, z_k)$, $k = 0,1,2$ are three points in the plane which are not co-linear.

The point at which the line intersects the plane is therefore described by setting the point on the line equal to the point on the plane, giving the parametric equation:

$$\mathbf{l}_a + (\mathbf{l}_b - \mathbf{l}_a)t = \mathbf{p}_0 + (\mathbf{p}_1 - \mathbf{p}_0)u + (\mathbf{p}_2 - \mathbf{p}_0)v$$

This can be rewritten as

$$\mathbf{l}_a - \mathbf{p}_0 = (\mathbf{l}_a - \mathbf{l}_b)t + (\mathbf{p}_1 - \mathbf{p}_0)u + (\mathbf{p}_2 - \mathbf{p}_0)v,$$

which can be expressed in matrix form as:

$$\begin{bmatrix} x_a - x_0 \\ y_a - y_0 \\ z_a - z_0 \end{bmatrix} = \begin{bmatrix} x_a - x_b & x_1 - x_0 & x_2 - x_0 \\ y_a - y_b & y_1 - y_0 & y_2 - y_0 \\ z_a - z_b & z_1 - z_0 & z_2 - z_0 \end{bmatrix} \begin{bmatrix} t \\ u \\ v \end{bmatrix}$$

The point of intersection is then equal to

$$\mathbf{l}_a + (\mathbf{l}_b - \mathbf{l}_a)t$$

If the line is parallel to the plane then the vectors $\mathbf{l}_b - \mathbf{l}_a$, $\mathbf{p}_1 - \mathbf{p}_0$, and $\mathbf{p}_2 - \mathbf{p}_0$ will be linearly dependent and the matrix will be singular. This situation will also occur when the line lies in the plane.

If the solution satisfies the condition $t \in [0,1],$, then the intersection point is on the line between $\mathbf{l}_a$ and $\mathbf{l}_b$.

If the solution satisfies

$$u, v \in [0,1], \quad (u + v) \le 1,$$

then the intersection point is in the plane inside the triangle spanned by the three points $\mathbf{p}_0$, and $\mathbf{p}_2$.

This problem is typically solved by expressing it in matrix form, and inverting it:

$$\begin{bmatrix} t \\ u \\ v \end{bmatrix} = \begin{bmatrix} x_a - x_b & x_1 - x_0 & x_2 - x_0 \\ y_a - y_b & y_1 - y_0 & y_2 - y_0 \\ z_a - z_b & z_1 - z_0 & z_2 - z_0 \end{bmatrix}^{-1} \begin{bmatrix} x_a - x_0 \\ y_a - y_0 \\ z_a - z_0 \end{bmatrix}.$$

Uses

In the ray tracing method of computer graphics a surface can be represented as a set of pieces of planes. The intersection of a ray of light with each plane is used to produce an image of the surface. In vision-based 3D reconstruction, a subfield of.computer vision, depth values are commonly measured by so-called triangulation method, which finds the intersection between light plane and ray reflected toward camera.

The algorithm can be generalised to cover intersection with other planar figures, in particular, the intersection of a polyhedron with a line.

Line Integrals

In mathematics, a line integral is an integral where the function to be integrated is evaluated along a curve. The terms path integral, curve integral, and curvilinear integral are also used; contour integral as well, although that is typically reserved for line integrals in the complex plane.

The function to be integrated may be a scalar field or a vector field. The value of the line integral is the sum of values of the field at all points on the curve, weighted by some scalar function on the curve (commonly arc length or, for a vector field, the scalar product of the vector field with a differential vector in the curve). This weighting distinguishes the line integral from simpler integrals defined on intervals. Many simple formulae in physics (for example, $W = \mathrm{F} \cdot \mathrm{s}$) have natural continuous analogs in terms of line integrals ($W = \int_C \mathrm{F} \cdot \mathrm{ds}$). The line integral finds the work done on an object moving through an electric or gravitational field, for example.

Vector Calculus

In qualitative terms, a line integral in vector calculus can be thought of as a measure of the total effect of a given field along a given curve. More specifically, the line integral over a scalar field can be interpreted as the area under the field carved out by a particular curve. This can be visualized as the surface created by $z = f(x,y)$ and a curve C in the x-y plane. The line integral of f would be the area of the "curtain" created when the points of the surface that are directly over C are carved out.

Line Integral of a Scalar Field

For some scalar field $f : U \subseteq \mathrm{R}^n \to \mathrm{R}$, the line integral along a piecewise smooth curve $C \subset U$ is defined as

$$\int_C f\, ds = \int_a^b f(\mathbf{r}(t))\,|\,\mathbf{r}'(t)\,|dt.$$

where r: [a, b] $\to C$ is an arbitrary bijective parametrization of the curve C such that r(a) and r(b) give the endpoints of C and $a < b$.

The function f is called the integrand, the curve C is the domain of integration, and the symbol ds may be intuitively interpreted as an elementary arc length. Line integrals of scalar fields over a curve C do not depend on the chosen parametrization r of C.

Geometrically, when the scalar field f is defined over a plane (n=2), its graph is a surface $z=f(x,y)$ in space, and the line integral gives the (signed) cross-sectional area bounded by the curve C and the graph of f.

Derivation

For a line integral over a scalar field, the integral can be constructed from a Riemann sum using the above definitions of f, C and a parametrization r of C. This can be done by partitioning the interval $[a,b]$ into n sub-intervals $[t_{i-1}, t_i]$ of length $\Delta t = (b$ " $a)/n$, then $r(t_i)$ denotes some point, call it a sample point, on the curve C. We can use the set of sample points $\{r(t_i) : 1 \leq i \leq n\}$ to approximate the curve C by a polygonal path by introducing a straight line piece between each of the sample points $r(t_{i-1})$ and $r(t_i)$. We then label the distance between each of the sample points on the curve as Δs_i. The product of $f(r(t_i))$ and Δs_i can be associated with the signed area of a rectangle with a height and width of $f(r(t_i))$ and Δs_i respectively. Taking the limit of the sum of the terms as the length of the partitions approaches zero gives us I

$$I = \lim_{\Delta s_i \to 0} \sum_{i=1}^{n} f(\mathbf{r}(t_i))\Delta s_i.$$

We note that, by the mean value theorem, the distance between subsequent points on the curve, is

$$\Delta s_i = |\mathbf{r}(t_i + \Delta t) - \mathbf{r}(t_i)| \approx |\mathbf{r}'(t_i)| \Delta t.$$

Substituting this in the above Riemann sum yields

$$I = \lim_{\Delta t \to 0} \sum_{i=1}^{n} f(\mathbf{r}(t_i)) |\mathbf{r}'(t_i)| \Delta t$$

which is the Riemann sum for the integral

$$I = \int_a^b f(\mathbf{r}(t)) |\mathbf{r}'(t)| dt.$$

Line integral of a vector field

Definition

For a vector field $F : U \subseteq R^n \to R^n$, the line integral along a piecewise smooth curve $C \subset U$, in the direction of r, is defined as

$$\int_C \mathbf{F}(\mathbf{r})\cdot d\mathbf{r} = \int_a^b \mathbf{F}(\mathbf{r}(t))\cdot\mathbf{r}'(t)dt.$$

where $\cdot$ is the dot product and r: [a, b] $\rightarrow C$ is a bijective parametrization of the curve C such that r(a) and r(b) give the endpoints of C.

A line integral of a scalar field is thus a line integral of a vector field where the vectors are always tangential to the line.

Line integrals of vector fields are independent of the parametrization r in absolute value, but they do depend on its orientation. Specifically, a reversal in the orientation of the parametrization changes the sign of the line integral.

The line integral of a vector field along a curve is the integral of the corresponding 1-form under the musical isomorphism over the curve considered as an immersed 1-manifold.

The line integral of a vector field can be derived in a manner very similar to the case of a scalar field. Again using the above definitions of F, C and its parametrization r(t), we construct the integral from a Riemann sum. Partition the interval $[a,b]$ into n intervals of length $\Delta t = (b - a)/n$. Letting t_i be the ith point on $[a,b]$, then r(t_i) gives us the position of the ith point on the curve. However, instead of calculating up the distances between subsequent points, we need to calculate their displacement vectors, Δr_i. As before, evaluating F at all the points on the curve and taking the dot product with each displacement vector gives us the infinitesimal contribution of each partition of F on C. Letting the size of the partitions go to zero gives us a sum

$$I = \lim_{\Delta t \to 0} \sum_{i=1}^{n} \mathbf{F}(\mathbf{r}(t_i)) \cdot \Delta \mathbf{r}_i$$

By the mean value theorem, we see that the displacement vector between adjacent points on the curve is

$$\Delta \mathbf{r}_i = \mathbf{r}(t_i + \Delta t) - \mathbf{r}(t_i) \approx \mathbf{r}'(t_i)\Delta t$$

Substituting this in the above Riemann sum yields

$$I = \lim_{\Delta t \to 0} \sum_{i=1}^{n} \mathbf{F}(\mathbf{r}(t_i)) \cdot \mathbf{r}'(t_i)\Delta t$$

which is the Riemann sum for the integral defined above.

Path Independence

If a vector field F is the gradient of a scalar field G (i.e. if F is conservative), that is,

$$\nabla G = \mathbf{F},$$

then the derivative of the composition of G and r(t) is

$$\frac{dG(\mathbf{r}(t))}{dt} = \nabla G(\mathbf{r}(t))\cdot\mathbf{r}'(t) = \mathbf{F}(\mathbf{r}(t))\cdot\mathbf{r}'(t)$$

which happens to be the integrand for the line integral of F on r(t). It follows that, given a path C, then

$$\int_C \mathbf{F}(\mathbf{r})\cdot d\mathbf{r} = \int_a^b \mathbf{F}(\mathbf{r}(t))\cdot\mathbf{r}'(t)dt = \int_a^b \frac{dG(\mathbf{r}(t))}{dt}dt = G(\mathbf{r}(b)) - G(\mathbf{r}(a)).$$

In other words, the integral of F over C depends solely on the values of G in the points r(b) and r(a) and is thus independent of the path between them.

For this reason, a line integral of a conservative vector field is called *path independent*.

Applications

The line integral has many uses in physics. For example, the work done on a particle travelling on a curve C inside a force field represented as a vector field F is the line integral of F on C.

Flow Across a Curve

For a vector field F : $U \subseteq \mathbb{R}^2 \to \mathbb{R}^2$, such as $F(x,y) = (P(x,y), Q(x,y))$ the line integral across a piecewise smooth curve $C \subset U$, is defined as

$$\int_C \mathbf{F}(\mathbf{r})\cdot d\mathbf{r}^{\perp} = \int_a^b -Q(x,y)dx + P(x,y)dy = \int_a^b (P(\mathbf{r}(t)), Q(\mathbf{r}(t)))\cdot(r_2'(t), -r_1'(t))dt.$$

where $\cdot$ is the dot product and r: [a, b] $\to C$, $r(t) = (r_1(t), r_2(t))$ is a bijective parametrization of the curve C such that r(a) and r(b) give the endpoints of C.

Complex Line Integral

In complex analysis, the line integral is defined in terms of multiplication and addition of complex numbers. Suppose U is an open subset of the complex plane C, $f : U \to$ C is a function, and $L \subset U$ is a curve of finite length, parametrized by $\gamma:[a,b] \to L$, where $\gamma(t) = x(t) + iy(t)$. The line integral

$$\int_L f(z)dz$$

may be defined by subdividing the interval [a, b] into $a = t_0 < t_1 < \ldots < t_n = b$ and considering the expression

$$\sum_{k=1}^{n} f(\gamma(t_k))[\gamma(t_k) - \gamma(t_{k-1})] = \sum_{k=1}^{n} f(\gamma_k)\Delta\gamma_k.$$

The integral is then the limit of this Riemann sum as the lengths of the subdivision intervals approach zero.

If the parametrization γ is continuously differentiable, the line integral can be evaluated as an integral of a function of a real variable:

When L is a closed curve, that is, its initial and final points coincide, the notation

$$\oint_L f(z)dz$$

is often used for the line integral of f along L. A closed curve line integral is sometimes referred to as a cyclic integral in engineering applications.

The line integral with respect to the conjugate complex differential $\overline{dz}$ is defined to be

$$\int_L f\overline{dz} := \overline{\int_L \overline{f}dz} = \int_a^b f(\gamma(t))\overline{\gamma'(t)}dt.$$

The line integrals of complex functions can be evaluated using a number of techniques: the integral may be split into real and imaginary parts reducing the problem to that of evaluating two real-valued line integrals, the Cauchy integral formula may be used in other circumstances. If the line integral is a closed curve in a region where the function is analytic and containing no singularities, then the value of the integral is simply zero; this is a consequence of the Cauchy integral theorem. The residue theorem allows contour integrals to be used in the complex plane to find integrals of real-valued functions of a real variable.

Example

Consider the function $f(z)=1/z$, and let the contour L be the unit circle about 0, parametrized by $\gamma(t)=e^{it}$ with t in $[0, 2\pi]$ (which generates the circle counterclockwise). Substituting, we find

$$\begin{aligned}\oint_L f(z)dz &= \int_0^{2\pi} \frac{1}{e^{it}} ie^{it}\,dt = i\int_0^{2\pi} e^{-it}e^{it}\,dt \\ &= i\int_0^{2\pi} dt = i(2\pi - 0) = 2\pi i.\end{aligned}$$

Here we have used the fact that any complex number z can be written as re^{it} where r is the modulus of z. On the unit circle this is fixed to 1, so the only variable left is the angle, which is denoted by t. This answer can be also verified by the Cauchy integral formula.

Relation between the Line Integral of a Vector Field and the Complex Line Integral

Viewing complex numbers as 2-dimensional vectors, the line integral of a 2-dimensional vector field corresponds to the real part of the line integral of the conjugate of the corresponding complex function of a complex variable. More specifically, if $\mathbf{r}(t) = (x(t), y(t))$ is a parameterization of L and $f(z) = u(z) + iv(z)$, then:

$$\int_L \overline{f(z)}dz = \int_L \bar{f}\,dx + i\int_L \bar{f}\,dy = \int_L (u,v)\cdot d\mathbf{r} + i\int_L (-v,u)\cdot d\mathbf{r},$$

provided that both integrals on the right hand side exist, and that the parametrization γ of L has the same orientation as $\mathbf{r}$ (just expand the Riemann sum for the lefthand integral and take the limit).

By Green's theorem, the area of a region enclosed by a smooth, closed, positively oriented curve is given by the integral

$$\frac{1}{2i}\int_L \bar{z}\;dz$$

This fact is used, for example, in the proof of the area theorem.

Due to the Cauchy-Riemann equations the curl of the vector field corresponding to the conjugate of a holomorphic function is zero. This relates through Stokes' theorem both types of line integral being zero.

Quantum Mechanics

The "path integral formulation" of quantum mechanics actually refers not to path integrals in this sense but to functional integrals, that is, integrals over a space of paths, of a function *of* a possible path. However, path integrals in the sense of this article are important in quantum mechanics; for example, complex contour integration is often used in evaluating probability amplitudes in quantum scattering theory.

Irrotational Fields

A vector field v is said to be *irrotational* if its curl is zero. That is, if

$$\nabla \times \mathbf{v} = \mathbf{0}.$$

For this reason, such vector fields are sometimes referred to as curl free field (curl-free vector field) or *curl-less* vector fields.

It is an identity of vector calculus that for any scalar field φ:

$$\nabla \times \nabla\varphi = \mathbf{0}.$$

Therefore every conservative vector field is also an irrotational vector field.

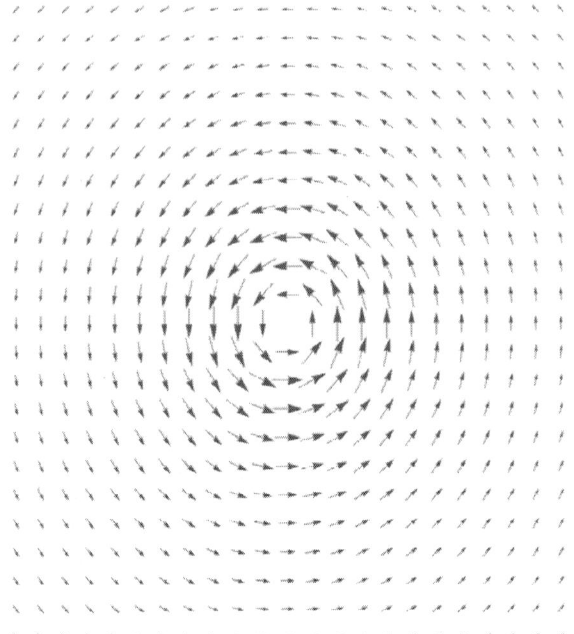

Figure: *The above field $v = (-y/(x^2 + y^2), +x/(x^2 + y^2), 0)$ includes a vortex at its center, so it is* non-*irrotational; it is neither conservative, nor does it have path independence. However, any simply connected subset that excludes the vortex line (0,0,z) will have zero curl, $\nabla \times v = 0$. Such vortex-free regions are examples of irrotational vector fields.*

Provided that S is a simply connected region, the converse of this is true: every irrotational vector field is also a conservative vector field.

The above statement S is *not* true if is not simply connected. Let S be the usual 3-dimensional space, except with the z-axis removed; that is $S = \mathbb{R}^3 \setminus \{(0,0,z) \mid z \in \mathbb{R}\}$. Now define a vector field by

$$\mathbf{v} = \left(\frac{-y}{x^2+y^2}, \frac{x}{x^2+y^2}, 0 \right).$$

Then $\mathbf{v}$ exists and has zero curl at every point in S; that is $\mathbf{v}$ is irrotational. However the circulation of $\mathbf{v}$ around the unit circle in the x, y-plane is equal to 2π. Indeed we note that in polar coordinates $\mathbf{v} = \mathbf{e}_\phi / r$, so the integral over the unit circle is equal $\int \mathbf{v}\mathbf{e}_\phi \mathrm{d}\phi = 2\pi$. Therefore $\mathbf{v}$ does not have the path independence property discussed above, and is not conservative. (However, in any simply connected subregion of S, it is still true that it is conservative. In fact, the field above is the gradient of $\arg(x+iy)$. As we know from complex analysis, this is a multi-valued function which requires a branch cut from the origin to infinity to be defined in a continuous way; hence, in a region that does not go around the z-axis, its gradient is conservative.)

In a simply connected region an irrotational vector field has the path independence property. This can be seen by noting that in such a region an irrotational vector field is conservative, and conservative vector fields have the path independence property. The result can also be proved directly by using Stokes' theorem. In a connected region any vector field which has the path independence property must also be irrotational.

More abstractly, a conservative vector field is an exact 1-form. That is, it is a 1-form equal to the exterior derivative of some 0-form (scalar field) ϕ. An irrotational vector field is a closed 1-form. Since $d^2 = 0$, any exact form is closed, so any conservative vector field is irrotational. The domain is simply connected if and only if its first homology group is 0, which is equivalent to its first cohomology group being 0. The first de Rham cohomology group H^1_{dR} is 0 if and only if all closed 1-forms are exact.

Curvilinear Coordinates

In geometry, curvilinear coordinates are a coordinate system for Euclidean space in which the coordinate lines may be curved. These coordinates may be derived from a set of Cartesian coordinates by using a transformation that is locally invertible (a one-to-one map) at each point. This means that one can convert a point given in a Cartesian coordinate system to its curvilinear coordinates and back. The name *curvilinear coordinates*, coined by the French mathematician Lamé, derives from the fact that the coordinate surfaces of the curvilinear systems are curved.

Well-known examples of curvilinear coordinate systems in three-dimensional Euclidean space (R^3) are Cartesian, cylindrical and spherical polar coordinates. A Cartesian coordinate surface in this space is a plane; for example $z = 0$ defines the x-y plane. In the same space, the coordinate surface $r = 1$ in spherical polar coordinates is the surface of a unit sphere, which is curved. The formalism of curvilinear coordinates provides a unified and general description of the standard coordinate systems.

Curvilinear coordinates are often used to define the location or distribution of physical quantities which may be, for example, scalars, vectors, or tensors. Mathematical expressions involving these quantities in vector calculus and tensor analysis (such as the gradient, divergence, curl, and Laplacian) can be transformed from one coordinate system to another, according to transformation rules for scalars, vectors, and tensors. Such expressions then become valid for any curvilinear coordinate system.

Depending on the application, a curvilinear coordinate system may be simpler to use than the Cartesian coordinate system. For instance, a physical problem with spherical symmetry defined in R^3 (for example, motion of particles under the influence of central forces) is usually easier to solve in spherical polar coordinates than in Cartesian coordinates.

Equations with boundary conditions that follow coordinate surfaces for a particular curvilinear coordinate system may be easier to solve in that system. One would for instance describe the motion of a particle in a rectangular box in Cartesian coordinates, whereas one would prefer spherical coordinates for a particle in a sphere. Spherical coordinates are one of the most used curvilinear coordinate systems in such fields as Earth sciences, cartography, and physics (in particular quantum mechanics, relativity), and engineering.

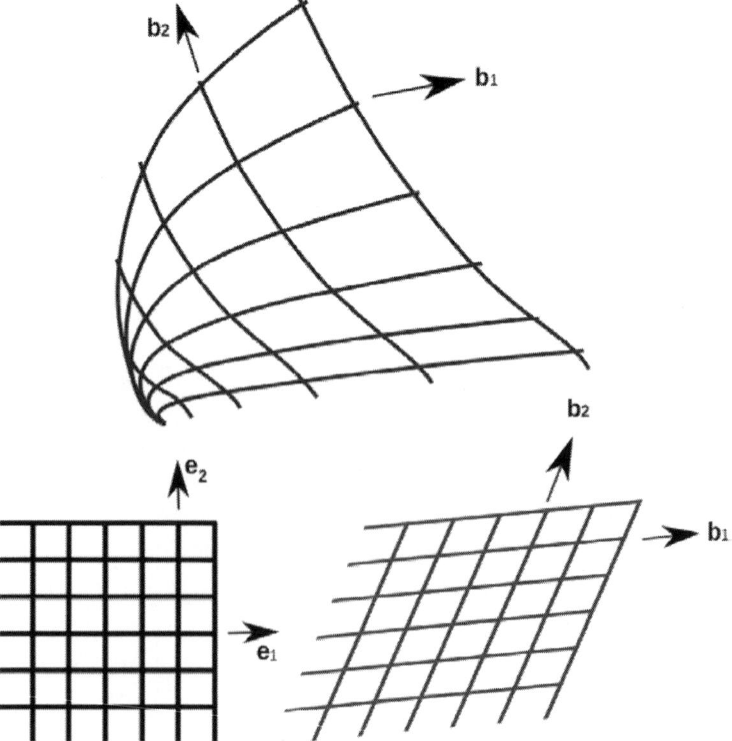

Figure: *Curvilinear, affine, and Cartesian coordinates in two-dimensional space*

Orthogonal Curvilinear Coordinates in 3d

Coordinates, Basis, and Vectors

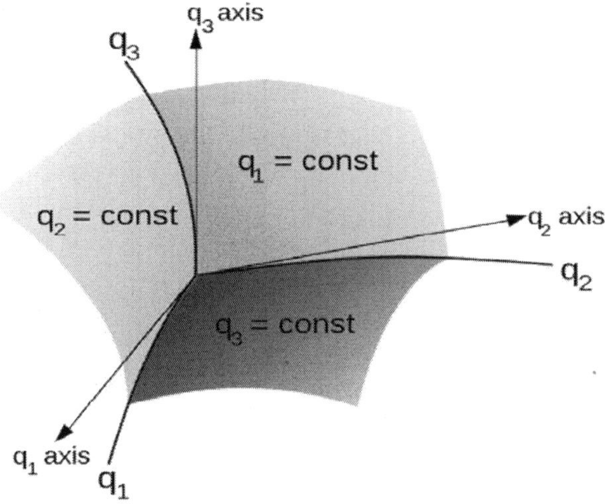

Figure: *Coordinate surfaces, coordinate lines, and coordinate axes of general curvilinear coordinates.*

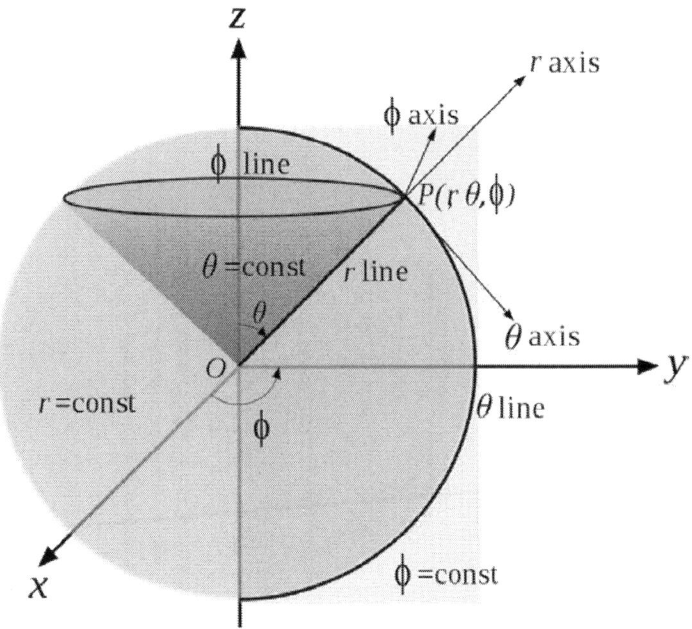

Figure: *Coordinate surfaces, coordinate lines, and coordinate axes of spherical coordinates. Surfaces:* r*- spheres,* θ *- cones,* φ *- half-planes; Lines:* r *- straight beams,* θ *- vertical semicircles,* φ *- horizontal circles; Axes:* r *- straight beams,* θ *- tangents to vertical semicircles,* φ *- tangents to horizontal circles*

For now, consider 3d space. A point P in 3d space can be defined using Cartesian coordinates (x, y, z) [equivalently written (x_1, x_2, x_3)], or in another system (q_1, q_2, q_3), as shown in Fig. 1. The latter is a curvilinear coordinate system, and (q_1, q_2, q_3) are the curvilinear coordinates of the point P.

The surfaces q_1 = constant, q_2 = constant, q_3 = constant are called the coordinate surfaces; and the space curves formed by their intersection in pairs are called the coordinate curves. The coordinate axes are determined by the tangents to the coordinate curves at the intersection of three surfaces. They are not in general fixed directions in space, which happens to be the case for simple Cartesian coordinates.

A basis whose vectors change their direction and/or magnitude from point to point is called local basis. All bases associated with curvilinear coordinates are necessarily local. Basis vectors that are the same at all points are global bases, and can be associated only with linear or affine coordinate systems.

Note: usually all basis vectors are denoted by e, for this article e is for the standard basis (Cartesian) and b is for the curvilinear basis.

The relation between the coordinates is given by the invertible transformations:

$$x = x(q_1, q_2, q_3), y = y(q_1, q_2, q_3), z = z(q_1, q_2, q_3)$$

$$q_1 = q_1(x, y, z), q_2 = q_2(x, y, z), q_3 = q_3(x, y, z)$$

Any point can be written as a position vector r in Cartesian coordinates:

$$\mathbf{r} = x\mathbf{e}_x + y\mathbf{e}_y + z\mathbf{e}_z$$

where x, y, z are the coordinates of the position vector with respect to the standard basis *vectors* e_x, e_y, e_z.

However, in a general curvilinear system, there may well not be any natural global basis vectors. Instead, we note that in the Cartesian system, we have the property that

$$\mathbf{e}_x = \frac{\partial \mathbf{r}}{\partial x}; \mathbf{e}_y = \frac{\partial \mathbf{r}}{\partial y}; \mathbf{e}_z = \frac{\partial \mathbf{r}}{\partial z}.$$

We can apply the same idea to the curvilinear system to determine a system of basis vectors at P. We define

$$\mathbf{h}_1 = \frac{\partial \mathbf{r}}{\partial q_1}; \mathbf{h}_2 = \frac{\partial \mathbf{r}}{\partial q_2}; \mathbf{h}_3 = \frac{\partial \mathbf{r}}{\partial q_3}.$$

These may not have unit length, and may also not be orthogonal. In the case that they *are* orthogonal at all points where the derivatives are well-defined, we define the Lamé coefficients (after Gabriel Lamé) by

$$h_1 = |\mathbf{h}_1|;\ h_2 = |\mathbf{h}_2|;\ h_3 = |\mathbf{h}_3|$$

and the curvilinear orthonormal basis vectors by

$$\mathbf{b}_1 = \frac{\mathbf{h}_1}{h_1};\ \mathbf{b}_2 = \frac{\mathbf{h}_2}{h_2};\ \mathbf{b}_3 = \frac{\mathbf{h}_3}{h_3}.$$

It is important to note that these basis vectors may well depend upon the position of *P*; it is therefore necessary that they are not assumed to be constant over a region. (They technically form a basis for the tangent bundle of $\mathbb{R}^3$ at *P*, and so are local to *P*.)

In general, curvilinear coordinates allow the generality of basis vectors not all mutually perpendicular to each other, and not required to be of unit length: they can be of arbitrary magnitude and direction. The use of an orthogonal basis makes vector manipulations simpler than for non-orthogonal. However, some areas of physics and engineering, particularly fluid mechanics and continuum mechanics, require non-orthogonal bases to describe deformations and fluid transport to account for complicated directional dependences of physical quantities. A discussion of the general case appears later on this page.

Differential Elements

Since the total differential change in r is

$$d\mathbf{r} = \frac{\partial \mathbf{r}}{\partial q_1} dq_1 + \frac{\partial \mathbf{r}}{\partial q_2} dq_2 + \frac{\partial \mathbf{r}}{\partial q_3} dq_3 = h_1 dq_1 \mathbf{b}_1 + h_2 dq_2 \mathbf{b}_2 + h_3 dq_3 \mathbf{b}_3$$

so scale factors are

$$h_i = \left| \frac{\partial \mathbf{r}}{\partial q_i} \right|$$

They can also be written for each component of r:

$$h^i{}_k = \frac{\partial x^i}{\partial q^k}.$$

However, this designation is very rarely used, largely replaced with the components of the metric tensor g_{ik}.

Covariant and Contravariant Bases

The basis vectors, gradients, and scale factors are all interrelated within a coordinate system by two methods:

1. the basis vectors are unit tangent vectors along the coordinate curves:

$$\mathbf{b}_i = \frac{1}{\left|\frac{\partial \mathbf{r}}{\partial q_i}\right|}\frac{\partial \mathbf{r}}{\partial q_i} = \frac{1}{h_i}\frac{\partial \mathbf{r}}{\partial q_i}$$

which transform like covariant vectors (denoted by lowered indices), or

2. the basis vectors are unit normal vectors to the coordinate surfaces:

$$\mathbf{b}^i = \frac{\nabla q_i}{|\nabla q_i|} = h_i \nabla q_i$$

which transform like contravariant vectors (denoted by raised indices), ∇ is the del operator.

So depending on the method by which they are built, for a general curvilinear coordinate system there are two sets of basis vectors for every point: $\{b_1, b_2, b_3\}$ is the covariant basis, and $\{b^1, b^2, b^3\}$ is the contravariant basis.

A vector v can be given in terms either basis, i.e.,

$$\mathbf{v} = v^1\mathbf{b}_1 + v^2\mathbf{b}_2 + v^3\mathbf{b}_3 = v_1\mathbf{b}^1 + v_2\mathbf{b}^2 + v_3\mathbf{b}^3$$

The basis vectors relate to the components by[2](pp30–32)

$$\mathbf{v}\cdot\mathbf{b}^i = v^k\mathbf{b}_k\cdot\mathbf{b}^i = v^k\delta^i_k = v^i$$

$$\mathbf{v}\cdot\mathbf{b}_i = v_k\mathbf{b}^k\cdot\mathbf{b}_i = v_k\delta^k_i = v_i$$

and

$$\mathbf{v}\cdot\mathbf{b}_i = v^k\mathbf{b}_k\cdot\mathbf{b}_i = g_{ki}v^k$$

$$\mathbf{v}\cdot\mathbf{b}^i = v_k\mathbf{b}^k\cdot\mathbf{b}^i = g^{ki}v_k$$

where g is the metric tensor.

A vector is covariant or contravariant if, respectively, its components are covariant (lowered indices, written v_k) or contravariant (raised indices, written v^k). From the above vector sums, it can be seen that contravariant vectors are represented with covariant basis vectors, and covariant vectors are represented with contravariant basis vectors.

A key convention in the representation of vectors and tensors in terms of indexed components and basis vectors is *invariance* in the sense that vector components which transform in a covariant manner (or contravariant manner) are paired with basis vectors that transform in a contravariant manner (or covariant manner).

Constructing a Covariant Basis in One Dimension

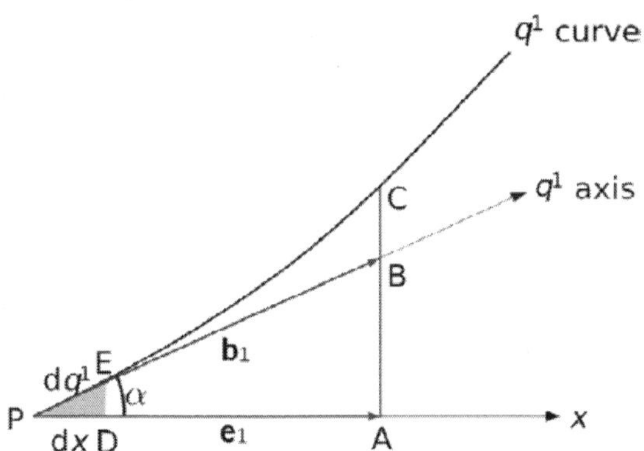

Figure: *Transformation of local covariant basis in the case of general curvilinear coordinates*

Consider the one-dimensional curve shown in Figure. At point *P*, taken as an origin, *x* is one of the Cartesian coordinates, and q^1 is one of the curvilinear coordinates. The local (non-unit) basis vector is b_1 (notated h_1 above, with b reserved for unit vectors) and it is built on the q^1 axis which is a tangent to that coordinate line at the point *P*. The axis q^1 and thus the vector b_1 form an angle *á* with the Cartesian *x* axis and the Cartesian basis vector e_1.

It can be seen from triangle *PAB* that

$$\cos\alpha = \frac{|\mathbf{e}_1|}{|\mathbf{b}_1|} \quad \Rightarrow \quad |\mathbf{e}_1| = |\mathbf{b}_1| \cos\alpha$$

where $|e_1|$, $|b_1|$ are the magnitudes of the two basis vectors, i.e., the scalar intercepts *PB* and *PA*. Note that *PA* is also the projection of b_1 on the *x* axis.

However, this method for basis vector transformations using *directional cosines* is inapplicable to curvilinear coordinates for the following reasons:

1. By increasing the distance from *P*, the angle between the curved ıın • q^1 and Cartesian axis *x* increasingly deviates from α.
2. At the distance *PB* the true angle is that which the tangent at point C forms with the *x* axis and the latter angle is clearly different from α.

The angles that the q^1 line and that axis form with the *x* axis become closer in value the closer one moves towards point *P* and become exactly equal at *P*.

Let point E be located very close to P, so close that the distance PE is infinitesimally small. Then PE measured on the q^1 axis almost coincides with PE measured on the q^1 line. At the same time, the ratio PD/PE (PD being the projection of PE on the x axis) becomes almost exactly equal to cos α.

Let the infinitesimally small intercepts PD and PE be labelled, respectively, as dx and dq^1. Then

$$\cos\alpha = \frac{dx}{dq^1} = \frac{|\mathbf{e}_1|}{|\mathbf{b}_1|}$$

Thus, the directional cosines can be substituted in transformations with the more exact ratios between infinitesimally small coordinate intercepts. It follows that the component (projection) of b_1 on the x axis is

$$p^1 = \mathbf{b}_1 \cdot \frac{\mathbf{e}_1}{|\mathbf{e}_1|} = |\mathbf{b}_1| \frac{|\mathbf{e}_1|}{|\mathbf{e}_1|} \cos\alpha = |\mathbf{b}_1| \frac{dx}{dq^1} \quad \Rightarrow \quad \frac{p^1}{|\mathbf{b}_1|} = \frac{dx}{dq^1}.$$

If $q^i = q^i(x_1, x_2, x_3)$ and $x_i = x_i(q^1, q^2, q^3)$ are smooth (continuously differentiable) functions the transformation ratios can be written as $\frac{\partial q^i}{\partial x_j}$ and $\frac{\partial x_i}{\partial q^j}$. That is, those ratios are partial derivatives of coordinates belonging to one system with respect to coordinates belonging to the other system.

Constructing a Covariant Basis in Three Dimensions

Doing the same for the coordinates in the other 2 dimensions, b_1 can be expressed as:

$$\mathbf{b}_1 = p^1\mathbf{e}_1 + p^2\mathbf{e}_2 + p^3\mathbf{e}_3 = \frac{\partial x_1}{\partial q^1}\mathbf{e}_1 + \frac{\partial x_2}{\partial q^1}\mathbf{e}_2 + \frac{\partial x_3}{\partial q^1}\mathbf{e}_3$$

Similar equations hold for b_2 and b_3 so that the standard basis $\{e_1, e_2, e_3\}$ is transformed to a local (ordered and *normalised*) basis $\{b_1, b_2, b_3\}$ by the following system of equations:

$$\mathbf{b}_1 = \frac{\partial x_1}{\partial q^1}\mathbf{e}_1 + \frac{\partial x_2}{\partial q^1}\mathbf{e}_2 + \frac{\partial x_3}{\partial q^1}\mathbf{e}_3$$

$$\mathbf{b}_2 = \frac{\partial x_1}{\partial q^2}\mathbf{e}_1 + \frac{\partial x_2}{\partial q^2}\mathbf{e}_2 + \frac{\partial x_3}{\partial q^2}\mathbf{e}_3$$

$$\mathbf{b}_3 = \frac{\partial x_1}{\partial q^3}\mathbf{e}_1 + \frac{\partial x_2}{\partial q^3}\mathbf{e}_2 + \frac{\partial x_3}{\partial q^3}\mathbf{e}_3$$

By analogous reasoning, one can obtain the inverse transformation from local basis to standard basis:

$$\mathbf{e}_1 = \frac{\partial q^1}{\partial x_1}\mathbf{b}_1 + \frac{\partial q^2}{\partial x_1}\mathbf{b}_2 + \frac{\partial q^3}{\partial x_1}\mathbf{b}_3$$

$$\mathbf{e}_2 = \frac{\partial q^1}{\partial x_2}\mathbf{b}_1 + \frac{\partial q^2}{\partial x_2}\mathbf{b}_2 + \frac{\partial q^3}{\partial x_2}\mathbf{b}_3$$

$$\mathbf{e}_3 = \frac{\partial q^1}{\partial x_3}\mathbf{b}_1 + \frac{\partial q^2}{\partial x_3}\mathbf{b}_2 + \frac{\partial q^3}{\partial x_3}\mathbf{b}_3$$

Jacobian of the Transformation

The above systems of linear equations can be written in matrix form as

$$\frac{\partial x_i}{\partial q^k}\mathbf{e}_i = \mathbf{b}_k, \quad \frac{\partial q^i}{\partial x_k}\mathbf{b}_i = \mathbf{e}_k .$$

This coefficient matrix of the linear system is the Jacobian matrix (and its inverse) of the transformation. These are the equations that can be used to transform a Cartesian basis into a curvilinear basis, and vice versa.

In three dimensions, the expanded forms of these matrices are

$$\mathbf{J} = \begin{bmatrix} \frac{\partial x_1}{\partial q^1} & \frac{\partial x_1}{\partial q^2} & \frac{\partial x_1}{\partial q^3} \\ \frac{\partial x_2}{\partial q^1} & \frac{\partial x_2}{\partial q^2} & \frac{\partial x_2}{\partial q^3} \\ \frac{\partial x_3}{\partial q^1} & \frac{\partial x_3}{\partial q^2} & \frac{\partial x_3}{\partial q^3} \end{bmatrix}, \quad \mathbf{J}^{-1} = \begin{bmatrix} \frac{\partial q^1}{\partial x_1} & \frac{\partial q^1}{\partial x_2} & \frac{\partial q^1}{\partial x_3} \\ \frac{\partial q^2}{\partial x_1} & \frac{\partial q^2}{\partial x_2} & \frac{\partial q^2}{\partial x_3} \\ \frac{\partial q^3}{\partial x_1} & \frac{\partial q^3}{\partial x_2} & \frac{\partial q^3}{\partial x_3} \end{bmatrix}$$

In the inverse transformation (second equation system), the unknowns are the curvilinear basis vectors. For all points there can only exist one and only one set of basis vectors (else vectors are not well defined at those points). This condition is satisfied if and only if the equation system has a single solution, from linear algebra, a linear equation system has a single solution (non-trivial) only if the determinant of its system matrix is non-zero:

$$\det(\mathbf{J}^{-1}) \neq 0$$

which shows the rationale behind the above requirement concerning the inverse Jacobian determinant.

Generalization to n Dimensions

The formalism extends to any finite dimension as follows.

Consider the real Euclidean n-dimensional space, that is $R^n = R \times R \times \ldots \times R$ (n times) where R is the set of real numbers and $\times$ denotes the Cartesian product, which is a vector space.

The coordinates of this space can be denoted by: $x = (x_1, x_2, \ldots, x_n)$. Since this is a vector (an element of the vector space), it can be written as:

$$\mathbf{x} = \sum_{i=1}^{n} x_i \mathbf{e}^i$$

where $e^1 = (1,0,0\ldots,0)$, $e^2 = (0,1,0\ldots,0)$, $e^3 = (0,0,1\ldots,0), \ldots, e^n = (0,0,0\ldots,1)$ is the standard basis *set of vectors* for the space R^n, and $i = 1, 2, \ldots n$ is an index labelling components. Each vector has exactly one component in each dimension (or "axis") and they are mutually orthogonal (perpendicular) and normalized (has unit magnitude).

More generally, we can define basis vectors b_i so that they depend on $q = (q_1, q_2, \ldots, q_n)$, i.e. they change from point to point: $b_i = b_i(q)$. In which case to define the same point x in terms of this alternative basis: the coordinates with respect to this basis v_i also necessarily depend on x also, that is $v_i = v_i(x)$. Then a vector v in this space, with respect to these alternative coordinates and basis vectors, can be expanded as a linear combination in this basis (which simply means to multiply each basis vector e_i by a number v_i – scalar multiplication):

$$\mathbf{v} = \sum_{j=1}^{n} \bar{v}^j \mathbf{b}_j = \sum_{j=1}^{n} \bar{v}^j(\mathbf{q})\mathbf{b}_j(\mathbf{q})$$

The vector sum that describes v in the new basis is composed of different vectors, although the sum itself remains the same.

Transformation of Coordinates

From a more general and abstract perspective, a curvilinear coordinate system is simply a coordinate patch on the differentiable manifold E^n (n-dimensional Euclidean space) that is diffeomorphic to the Cartesian coordinate patch on the manifold.[3] Note that two diffeomorphic coordinate patches on a differential manifold need not overlap differentiably. With this simple definition of a curvilinear coordinate system, all the results that follow below are simply applications of standard theorems in differential topology.

The transformation functions are such that there's a one-to-one relationship between points in the "old" and "new" coordinates, that is, those functions are bijections, and fulfil the following requirements within their domains:

1. They are smooth functions: $q^i = q^i(x)$
2. The inverse Jacobian determinant

$$J^{-1} = \begin{vmatrix} \frac{\partial q^1}{\partial x_1} & \frac{\partial q^1}{\partial x_2} & \cdots & \frac{\partial q^1}{\partial x_n} \\ \frac{\partial q^2}{\partial x_1} & \frac{\partial q^2}{\partial x_2} & \cdots & \frac{\partial q^2}{\partial x_n} \\ \vdots & \vdots & \ddots & \vdots \\ \frac{\partial q^n}{\partial x_1} & \frac{\partial q^n}{\partial x_2} & \cdots & \frac{\partial q^n}{\partial x_n} \end{vmatrix} \neq 0$$

is not zero; meaning the transformation is invertible: $x_i(q)$. *according to the inverse function theorem. The condition that the Jacobian determinant is not zero reflects the fact that three surfaces from different families intersect in one and only one point and thus determine the position of this point in a unique way.*

Vector and Tensor Algebra in Three-Dimensional Curvilinear Coordinates

Elementary vector and tensor algebra in curvilinear coordinates is used in some of the older scientific literature in mechanics and physics and can be indispensable to understanding work from the early and mid-1900s, for example the text by Green and Zerna. Some useful relations in the algebra of vectors and second-order tensors in curvilinear coordinates are given in this section. The notation and contents are primarily from Ogden, Naghdi, Simmonds, Green and Zerna, Basar and Weichert, and Ciarlet.

Tensors in Curvilinear Coordinates

A second-order tensor can be expressed as

$$\mathbf{S} = S^{ij}\mathbf{b}_i \otimes \mathbf{b}_j = S^i{}_j\mathbf{b}_i \otimes \mathbf{b}^j = S_i{}^j\mathbf{b}^i \otimes \mathbf{b}_j = S_{ij}\mathbf{b}^i \otimes \mathbf{b}^j$$

where $\otimes$ denotes the tensor product. The components S^{ij} are called the contravariant components, $S^i{}_j$ the mixed right-covariant components, $S_i{}^j$ the mixed left-covariant components, and S_{ij} the covariant components of the second-order tensor. The components of the second-order tensor are related by

$$S^{ij} = g^{ik} S_k^{\ j} = g^{jk} S^i_{\ k} = g^{ik} g^{j\ell} S_{k\ell}$$

The Metric Tensor in Orthogonal Curvilinear Coordinates

At each point, one can construct a small line element dx, so the square of the length of the line element is the scalar product dx • dx and is called the metric of the space, given by:

$$d\mathbf{x} \cdot d\mathbf{x} = \frac{\partial x_i}{\partial q^j} \frac{\partial x_i}{\partial q^k} dq^j dq^k$$

and the *symmetric* quantity

$$g_{ij}(q^i, q^j) = \frac{\partial x_k}{\partial q^i} \frac{\partial x_k}{\partial q^j} = \mathbf{b}_i \cdot \mathbf{b}_j$$

is called the fundamental (or metric) tensor of the Euclidean space in curvilinear coordinates.

Indices can be raised and lowered by the metric:

$$v^i = g^{ik} v_k$$

Relation to Lamé Coefficients

Defining the scale factors h_i by

$$h_i h_j = g_{ij} = \mathbf{b}_i \cdot \mathbf{b}_j \quad \Rightarrow \quad h_i = \sqrt{g_{ii}} = |\mathbf{b}_i| = \left| \frac{\partial \mathbf{x}}{\partial q^i} \right|$$

gives a relation between the metric tensor and the Lamé coefficients. Note also that

$$g_{ij} = \frac{\partial \mathbf{x}}{\partial q^i} \cdot \frac{\partial \mathbf{x}}{\partial q^j} = \left(h_{ki} \mathbf{e}_k \right) \cdot \left(h_{mj} \mathbf{e}_m \right) = h_{ki} h_{kj}$$

where h_{ij} are the Lamé coefficients. For an orthogonal basis we also have:

$$g = g_{11} g_{22} g_{33} = h_1^2 h_2^2 h_3^2 \quad \Rightarrow \quad \sqrt{g} = h_1 h_2 h_3 = J$$

Example: Polar Coordinates

If we consider polar coordinates for R^2, note that

$$(x, y) = (r \cos \theta, r \sin \theta)$$

(r, θ) are the curvilinear coordinates, and the Jacobian determinant of the transformation $(r,\theta) \to (r \cos \theta, r \sin \theta)$ is r.

The orthogonal basis vectors are $b_r = (\cos \theta, \sin \theta)$, $b_\theta = (-r \sin \theta, r \cos \theta)$. The normalized basis vectors are $e_r = (\cos \theta, \sin \theta)$, $e_\theta = (-\sin \theta, \cos \theta)$

and the scale factors are $h_r = 1$ and $h_\theta = r$. The fundamental tensor is $g_{11} = 1$, $g_{22} = r^2$, $g_{12} = g_{21} = 0$.

The Alternating Tensor

In an orthonormal right-handed basis, the third-order alternating tensor is defined as

$$\mathcal{E} = \epsilon_{ijk}\mathbf{e}^i \otimes \mathbf{e}^j \otimes \mathbf{e}^k$$

In a general curvilinear basis the same tensor may be expressed as

$$\mathcal{E} = \mathcal{E}_{ijk}\mathbf{b}^i \otimes \mathbf{b}^j \otimes \mathbf{b}^k = \mathcal{E}^{ijk}\mathbf{b}_i \otimes \mathbf{b}_j \otimes \mathbf{b}_k$$

It can also be shown that

$$\mathcal{E}^{ijk} = \frac{1}{J}\epsilon_{ijk} = \frac{1}{+\sqrt{g}}\epsilon_{ijk}$$

Christoffel Symbols

Christoffel symbols of the first kind

$$\mathbf{b}_{i,j} = \frac{\partial \mathbf{b}_i}{\partial q^j} = \Gamma_{ijk}\mathbf{b}^k \quad \Rightarrow \quad \mathbf{b}_{i,j} \cdot \mathbf{b}_k = \Gamma_{ijk}$$

where the comma denotes a partial derivative. To express Γ_{ijk} in terms of g_{ij} we note that

$$\begin{aligned}
g_{ij,k} &= (\mathbf{b}_i \cdot \mathbf{b}_j)_{,k} = \mathbf{b}_{i,k} \cdot \mathbf{b}_j + \mathbf{b}_i \cdot \mathbf{b}_{j,k} = \Gamma_{ikj} + \Gamma_{jki} \\
g_{ik,j} &= (\mathbf{b}_i \cdot \mathbf{b}_k)_{,j} = \mathbf{b}_{i,j} \cdot \mathbf{b}_k + \mathbf{b}_i \cdot \mathbf{b}_{k,j} = \Gamma_{ijk} + \Gamma_{kji} \\
g_{jk,i} &= (\mathbf{b}_j \cdot \mathbf{b}_k)_{,i} = \mathbf{b}_{j,i} \cdot \mathbf{b}_k + \mathbf{b}_j \cdot \mathbf{b}_{k,i} = \Gamma_{jik} + \Gamma_{kij}
\end{aligned}$$

Since

$$\mathbf{b}_{i,j} = \mathbf{b}_{j,i} \quad \Rightarrow \quad \Gamma_{ijk} = \Gamma_{jik}$$

using these to rearrange the above relations gives

$$\Gamma_{ijk} = \frac{1}{2}(g_{ik,j} + g_{jk,i} - g_{ij,k}) = \frac{1}{2}[(\mathbf{b}_i \cdot \mathbf{b}_k)_{,j} + (\mathbf{b}_j \cdot \mathbf{b}_k)_{,i} - (\mathbf{b}_i \cdot \mathbf{b}_j)_{,k}]$$

Christoffel symbols of the second kind

$$\Gamma_{ij}{}^k = \Gamma_{ji}{}^k, \quad \frac{\partial \mathbf{b}_i}{\partial q^j} = \Gamma_{ij}{}^k \mathbf{b}_k$$

This implies that

$$\Gamma_{ij}{}^k = \frac{\partial \mathbf{b}_i}{\partial q^j} \cdot \mathbf{b}^k = -\mathbf{b}_i \cdot \frac{\partial \mathbf{b}^k}{\partial q^j}$$

Other relations that follow are

$$\frac{\partial \mathbf{b}^i}{\partial q^j} = -\Gamma^i{}_{jk}\mathbf{b}^k, \quad \nabla \mathbf{b}_i = \Gamma_{ij}{}^k \mathbf{b}_k \otimes \mathbf{b}^j, \quad \nabla \mathbf{b}^i = -\Gamma_{jk}{}^i \mathbf{b}^k \otimes \mathbf{b}^j$$

Vector Operations

1. ***Dot Product:*** The scalar product of two vectors in curvilinear coordinates is $\mathbf{u}\cdot\mathbf{v} = u^i v_i = u_i v^i = g_{ij} u^i v^j = g^{ij} u_i v_j$
2. ***Cross Product:*** The cross product of two vectors is given by $\mathbf{u} \times \mathbf{v} = \epsilon_{ijk} u_j v_k \mathbf{e}_i$

where ϵ_{ijk} is the permutation symbol and $\mathbf{e}_i$ is a Cartesian basis vector. In curvilinear coordinates, the equivalent expression is

$$\mathbf{u} \times \mathbf{v} = [(\mathbf{b}_m \times \mathbf{b}_n)\cdot\mathbf{b}_s] u^m v^n \mathbf{b}^s = \mathcal{E}_{smn} u^m v^n \mathbf{b}^s$$

where $\mathcal{E}_{ijk}$ is the third-order alternating tensor.

Vector and Tensor Calculus in Three-Dimensional Curvilinear Coordinates

Adjustments need to be made in the calculation of line, surface and volume integrals. For simplicity, the following restricts to three dimensions and orthogonal curvilinear coordinates. However, the same arguments apply for n-dimensional spaces. When the coordinate system is not orthogonal, there are some additional terms in the expressions.

Simmonds, in his book on tensor analysis, quotes Albert Einstein saying

> *The magic of this theory will hardly fail to impose itself on anybody who has truly understood it; it represents a genuine triumph of the method of absolute differential calculus, founded by Gauss, Riemann, Ricci, and Levi-Civita.*

Vector and tensor calculus in general curvilinear coordinates is used in tensor analysis on four-dimensional curvilinear manifolds in general relativity, in the mechanics of curved shells, in examining the invariance properties of Maxwell's equations which has been of interest in metamaterials and in many other fields.

Some useful relations in the calculus of vectors and second-order tensors in curvilinear coordinates are given in this section. The notation and contents are primarily from Ogden, Simmonds, Green and Zerna, Basar and Weichert, and Ciarlet.

Let φ = φ(x) be a well defined scalar field and v = v(x) a well-defined vector field, and λ_1, λ_2... be parameters of the coordinates

Geometric Elements

1. Tangent vector: If x(λ) parametrizes a curve C in Cartesian coordinates, then

$$\frac{\partial \mathbf{x}}{\partial \lambda} = \frac{\partial \mathbf{x}}{\partial q^i}\frac{\partial q^i}{\partial \lambda} = \left(h_{ki}\frac{\partial q^i}{\partial \lambda}\right)\mathbf{b}_k$$

is a tangent vector to C in curvilinear coordinates (using the chain rule). Using the definition of the Lamé coefficients, and that for the metric $g_{ij} = 0$ when $i \neq j$, the magnitude is:

$$\left|\frac{\partial \mathbf{x}}{\partial \lambda}\right| = \sqrt{h_{ki}h_{kj}\frac{\partial q^i}{\partial \lambda}\frac{\partial q^j}{\partial \lambda}} = \sqrt{g_{ij}\frac{\partial q^i}{\partial \lambda}\frac{\partial q^j}{\partial \lambda}} = \sqrt{h_i^2\left(\frac{\partial q^i}{\partial \lambda}\right)^2}$$

2. Tangent plane element: If x(λ_1, λ_2) parametrizes a surface S in Cartesian coordinates, then the following cross product of tangent vectors is a normal vector to S with the magnitude of infinitesimal plane element, in curvilinear coordinates. Using the above result,

$$\frac{\partial \mathbf{x}}{\partial \lambda_1} \times \frac{\partial \mathbf{x}}{\partial \lambda_2} = \left(\frac{\partial \mathbf{x}}{\partial q^i}\frac{\partial q^i}{\partial \lambda_1}\right) \times \left(\frac{\partial \mathbf{x}}{\partial q^j}\frac{\partial q^j}{\partial \lambda_2}\right) = \mathcal{E}_{kmp}\left(h_{ki}\frac{\partial q^i}{\partial \lambda_1}\right)\left(h_{mj}\frac{\partial q^j}{\partial \lambda_2}\right)\mathbf{b}_p$$

where $\mathcal{E}$ is the permutation symbol. In determinant form:

$$\frac{\partial \mathbf{x}}{\partial \lambda_1} \times \frac{\partial \mathbf{x}}{\partial \lambda_2} = \begin{vmatrix} \mathbf{e}_1 & \mathbf{e}_2 & \mathbf{e}_3 \\ h_{1i}\frac{\partial q^i}{\partial \lambda_1} & h_{2i}\frac{\partial q^i}{\partial \lambda_1} & h_{3i}\frac{\partial q^i}{\partial \lambda_1} \\ h_{1j}\frac{\partial q^j}{\partial \lambda_2} & h_{2j}\frac{\partial q^j}{\partial \lambda_2} & h_{3j}\frac{\partial q^j}{\partial \lambda_2} \end{vmatrix}$$

Fictitious Forces in General Curvilinear Coordinates

An inertial coordinate system is defined as a system of space and time coordinates x_1, x_2, x_3, t in terms of which the equations of motion of a particle free of external forces are simply $d^2x_j/dt^2 = 0$. In this context, a coordinate system can fail to be "inertial" either due to non-straight time axis or non-straight space axes (or both). In other words, the basis vectors of the coordinates may vary in time at fixed positions, or they may vary with position at fixed times, or both. When equations of motion are expressed in terms of any non-inertial coordinate system (in this sense), extra terms appear, called Christoffel symbols.

Strictly speaking, these terms represent components of the absolute acceleration (in classical mechanics), but we may also choose to continue to regard d^2x_j/dt^2 as the acceleration (as if the coordinates were inertial) and treat the extra terms as if they were forces, in which case they are called fictitious forces. The component of any such fictitious force normal to the path of the particle and in the plane of the path's curvature is then called centrifugal force.

This more general context makes clear the correspondence between the concepts of centrifugal force in rotating coordinate systems and in stationary curvilinear coordinate systems. (Both of these concepts appear frequently in the literature.) For a simple example, consider a particle of mass m moving in a circle of radius r with angular speed w relative to a system of polar coordinates rotating with angular speed W. The radial equation of motion is $mr'' = F_r + mr(w + W)^2$.

Thus the centrifugal force is mr times the square of the absolute rotational speed $A = w + W$ of the particle. If we choose a coordinate system rotating at the speed of the particle, then $W = A$ and $w = 0$, in which case the centrifugal force is mrA^2, whereas if we choose a stationary coordinate system we have $W = 0$ and $w = A$, in which case the centrifugal force is again mrA^2. The reason for this equality of results is that in both cases the basis vectors at the particle's location are changing in time in exactly the same way. Hence these are really just two different ways of describing exactly the same thing, one description being in terms of rotating coordinates and the other being in terms of stationary curvilinear coordinates, both of which are non-inertial according to the more abstract meaning of that term.

When describing general motion, the actual forces acting on a particle are often referred to the instantaneous osculating circle tangent to the path of motion, and this circle in the general case is not centered at a fixed location, and so the decomposition into centrifugal and Coriolis components is constantly changing. This is true regardless of whether the motion is described in terms of stationary or rotating coordinates.

Tensor Algebra

In mathematics, the tensor algebra of a vector space V, denoted $T(V)$ or $T^{\cdot}(V)$, is the algebra of tensors on V (of any rank) with multiplication being the tensor product. It is the free algebra on V, in the sense of being left adjoint to the forgetful functor from algebras to vector spaces: it is the "most general" algebra containing V, in the sense of the corresponding universal property.

The tensor algebra also has two coalgebra structures; one simple one, which does not make it a bialgebra, and a more complicated one, which yields a bialgebra, and can be extended with an antipode to a Hopf algebra structure.

Construction

Let V be a vector space over a field K. For any nonnegative integer k, we define the kth tensor power of V to be the tensor product of V with itself k times:

$$T^k V = V^{\otimes k} = V \otimes V \otimes \cdots \otimes V.$$

That is, $T^k V$ consists of all tensors on V of rank k. By convention $T^0 V$ is the ground field K (as a one-dimensional vector space over itself).

We then construct $T(V)$ as the direct sum of $T^k V$ for $k = 0,1,2,\ldots$

$$T(V) = \bigoplus_{k=0}^{\infty} T^k V = K \oplus V \oplus (V \otimes V) \oplus (V \otimes V \otimes V) \oplus \cdots.$$

The multiplication in $T(V)$ is determined by the canonical isomorphism

$$T^k V \otimes T^\ell V \to T^{k+\ell} V$$

given by the tensor product, which is then extended by linearity to all of $T(V)$. This multiplication rule implies that the tensor algebra $T(V)$ is naturally a graded algebra with $T^k V$ serving as the grade-k subspace. This grading can be extended to a Z grading by appending subspaces $T^k V = \{0\}$ for negative integers k.

The construction generalizes in straightforward manner to the tensor algebra of any module M over a commutative ring. If R is a non-commutative ring, one can still perform the construction for any R-R bimodule M. (It does not work for ordinary R-modules because the iterated tensor products cannot be formed.)

Adjunction and Universal Property

The tensor algebra $T(V)$ is also called the free algebra on the vector space V, and is functorial. As with other free constructions, the functor T is left adjoint to some forgetful functor. In this case, it's the functor which sends each K-algebra to its underlying vector space.

Explicitly, the tensor algebra satisfies the following universal property, which formally expresses the statement that it is the most general algebra containing V:

Any linear transformation $f: V \rightarrow A$ from V to an algebra A over K can be uniquely extended to an algebra homomorphism from $T(V)$ to A as indicated by the following commutative diagram:

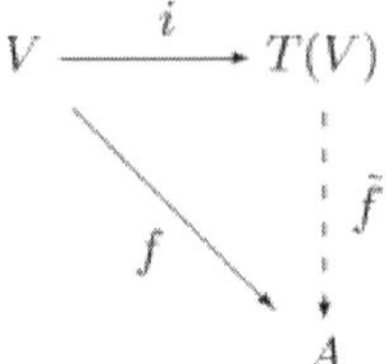

Here i is the canonical inclusion of V into $T(V)$ (the unit of the adjunction). One can, in fact, define the tensor algebra $T(V)$ as the unique algebra satisfying this property (specifically, it is unique up to a unique isomorphism), but one must still prove that an object satisfying this property exists.

The above universal property shows that the construction of the tensor algebra is *functorial* in nature. That is, T is a functor from the K-Vect, category of vector spaces over K, to K-Alg, the category of K-algebras. The functoriality of T means that any linear map from V to W extends uniquely to an algebra homomorphism from $T(V)$ to $T(W)$.

Non-Commutative Polynomials

If V has finite dimension n, another way of looking at the tensor algebra is as the "algebra of polynomials over K in n non-commuting variables". If we take basis vectors for V, those become non-commuting variables (or indeterminants) in $T(V)$, subject to no constraints beyond associativity, the distributive law and K-linearity.

Note that the algebra of polynomials on V is not $T(V)$, but rather $T(V^*)$: a (homogeneous) linear function on V is an element of V^*, for example coordinates $x^1,\ldots,x^n$ on a vector space are covectors, as they take in a vector and give out a scalar (the given coordinate of the vector).

Quotients

Because of the generality of the tensor algebra, many other algebras of interest can be constructed by starting with the tensor algebra and then imposing certain relations on the generators, i.e. by constructing certain quotient algebras of $T(V)$. Examples of this are the exterior algebra, the symmetric algebra, Clifford algebras and universal enveloping algebras.

Coalgebra Structures

The tensor algebra has two coalgebra structures; one simple one, which does not make it a bialgebra, and a more complicated one, which yields a bialgebra, and can be extended with an antipode to a Hopf algebra structure.

Simple Coalgebra Structure

The simple coalgebra structure on the tensor algebra is given as follows. The coproduct Δ is defined by

$$\Delta(v_1 \otimes \ldots \otimes v_m) := \sum_{i=0}^{m} (v_1 \otimes \ldots \otimes v_i) \otimes (v_{i+1} \otimes \ldots \otimes v_m)$$

extended by linearity to all of *TV*. The counit is given by

$$\varepsilon(v) = v \text{ for every } v \in T^0(V) \text{ and}$$

$$\varepsilon(v) = 0 \text{ for every } v \in T^k(V) \text{ for every } k > 0.$$

Note that $\Delta : TV \to TV \otimes TV$ respects the grading

$$T^m V \to \bigoplus_{i+j=m} T^i V \otimes T^j V$$

and ε is also compatible with the grading.

The tensor algebra is *not* a bialgebra with this coproduct.

Bialgebra and Hopf Algebra Structure

However, the following more complicated coproduct does yield a bialgebra:

$$\Delta(x_1 \otimes \ldots \otimes x_m) = \sum_{p=0}^{m} \sum_{\sigma \in \mathrm{Sh}_{p,m-p}} \left(x_{\sigma(1)} \otimes \ldots \otimes x_{\sigma(p)}\right) \otimes \left(x_{\sigma(p+1)} \otimes \ldots \otimes x_{\sigma(m)}\right)$$

where the summation is taken over all (p,m-p)-shuffles.

Finally, the tensor algebra becomes a Hopf algebra with antipode given by

$$S(x_1 \otimes \ldots \otimes x_m) = (-1)^m x_m \otimes \ldots \otimes x_1$$

extended linearly to all of *TV*.

This is just the standard Hopf algebra structure on a free algebra, where one defines the comultiplication on $T^1(V) = V$ by

$$\Delta(x) = x \otimes 1 + 1 \otimes x$$

and then extends to $T^m(V)$ via

$$\Delta(x_1 \otimes \ldots \otimes x_m) = \Delta(x_1)\Delta(x_2)\cdots\Delta(x_m).$$

Similarly one defines the antipode on $T^1(V) = V$ by

$$S(x) = -x$$

and then extends the antipode as the unique antiautomorphism of $T(V)$ with this property, i.e. we define the antipode on $T^m(V)$ via

$$S(x_1 \otimes \ldots \otimes x_m) = S(x_m)S(x_{m-1})\cdots S(x_2)S(x_1).$$

Bibliography

Albert, A.: *Regression and the Moore-Penrose Pseudoinverse,* Academic Press, New York, USA, 1972.

Aris, Rutherford, *Vectors, tensors, and the basic equations of fluid mechanics,* Dover, 1989.

Bishop, R. L. and R. J.: *Crittenden Geometry of Manifolds,* Academic Press, New York, USA, 1964.

Bonola, R.: *History of Non-Euclidean Geometry.* Hafner Publishing Company, New York, USA, 1912.

Boothby, W. M.: *An Introduction to Differentiable Manifolds and Riemannian Geometry,* Academic Press, New York, USA, 1975.

Boyd, John P., *Chebyshev and Fourier Spectral Methods,* Dover, 2000.

Carslaw, H.S.; and Jaeger, J.C., *Conduction of Heat in Solids*, Oxford University Press, 1959.

Cartan, E.: *Geometry of Riemannian Spaces,* Math Sci Press, Brookline, MA, USA, 1983.

Chris Doran and Anthony Lasenby, *Geometric Algebra for Physicists,* Cambridge University Press, Cambridge, 2003.

Crowe, M.J., *A History of Vector Analysis,* Notre Dame University Press, 1967.

Darling, R. W. R.: *Differential Forms and Connections,* Cambridge University Press, Cambridge, UK, 1994.

Dautray R. and Lions, J. L., *Spectral Theory and Applications,* Springer-Verlag, 1990.

Dennis Zill, G.; Michael Cullen R., *Advanced engineering mathematics,* Jones & Bartlett Learning, 2006.

Do Carmo, M. P.: *Differential Geometry of Curves and Surfaces,* Prentice Hall, Englewood Cliffs, NJ, USA, 1976.

Doolin, B. F. and C. F. Martin: *Introduction to Differential Geometry for Engineers,* Marcel Dekker, Inc., New York, NY, USA, 1990.

Dummit, David S.; Foote, Richard M., *Abstract Algebra*, John Wiley & Sons, 2004.

Eisenhart, L. P.: *An Introduction to Differential Geometry with Use of the Tensor Calculus,* Princeton University Press, Princeton, USA, 1940.

Farebrother, R. W.: *Linear Least Squares Computations*, Marcel Dekker Inc., New York, USA, 1988.

Field, Judith V.: *The Geometrical Work of Girard Desargues.* Springer-Verlag, New York, 1987.

George Arfken B. and Hans Weber, J., *Mathematical Methods for Physicists*, Academic Press, 1995.

Gerretsen, J. C. H.: *Lectures on Tensor Calculus and Differential Geometry,*Academic Press, New York, USA, 1982.

Girault V. and Raviart, P.A., *Finite Element Methods for Navier–Stokes Equations: Theory and Algorithms,* Springer-Verlag, 1986.

Gray, A.: *Modern Differential Geometry of Curves and Surfaces,* CRC Press, Boca Raton, FL, USA, 1993.

Gruber, M. H. J.: *Regression Estimators: A Comparitive Study*, Academic Press, San Diego CA, 1990.

Jacques Willy Delleur, *The handbook of groundwater engineering,* CRC Press, 2006.

Jeffreys, H. and Jeffreys, BS., *Methods of mathematical physics*, Cambridge University Press, 1999.

John Hopcroft E. and Jeffrey Ullman, D., *Introduction to Automata Theory, Languages and Computation*, Addison-Wesley Publishing, Reading Massachusetts, 1979.

Joseph, Pedlosky, *Geophysical fluid dynamics*, Springer, 1987.

Kendall, M. G.: *A Course in the Geometry of n Dimensions*, Hafner Publishing Company, New York, USA, 1961.

Lass, Harry, *Vector and Tensor Analysis,* McGraw-Hill Book Company, 1950.

Lay, David C., *Linear Algebra and Its Applications*, Wesley, 2006.

Levi-Civita, T.: *The Absolute Differential Calculus,* Dover Publications, New York, USA, 1977.

Lounesto, Pertti, *Clifford algebras and spinors*, Cambridge University Press, Cambridge, 2001.

Marsden, J.E., *Vector Calculus*, W. H. Freeman & Company, 1976.

Maxwell, James Clerk, *Treatise on Electricity and Magnetism*, 1892.

Michael Crowe J., *A History of Vector Analysis : The Evolution of the Idea of a Vectorial System*, Dover Publications, 1967.

Millman, R. S. and G. D. Parker: *Elements of Differential Geometry,*Dover Publications, New York, USA, 2009.

Mlodinow, M.: *Euclid's window (the story of geometry from parallel lines to hyperspace)*, Harper & Row, New York, USA, 1974.

Murray, M. K. and J. W. Rice: *Differential Geometry and Statistics*, Chapman & Hall, London, UK. 1993.

Nahin, Paul J., *Oliver Heaviside: the life, work, and times of an electrical genius of the Victorian age*, JHU Press, 2000.

Nomizu, K. and T. Sasaki: *Affine Differential Geometry*, Cambridge University Press, Cambridge, UK, 1966.

Pertti Lounesto, *Clifford algebras and spinors*, Cambridge University Press, 2001.

Postnikov, M. M.: *The Variational Theory of Geodesics*, Dover, Publications, New York, 1983.

Robinson, E. A.: *Least Squares Regression Analysis in Terms of Linear Algebra*, Goose Pond Press, Houston, USA, 1981.

Saville, D. J. and G. R. Wood: *Statistical Methods: The Geometric Approach*, Springer-Verlag, New York, USA, 1991.

Schey , H. M., *Div Grad Curl and all that: An informal text on vector calculus*, W. W. Norton & Company, 2005.

Serge, Lang, *Graduate Texts in Mathematics*, Springer, 2002.

Sheldon, Axler, *Linear Algebra Done Right*, Springer, 2002.

Strang, Gilbert, *Linear Algebra and Its Applications*, Brooks Cole, 2006.

Suthan Suthersan S., *Remediation engineering: design concepts*, CRC Press, 1996.

Weintraub, S. H.: *Differential Forms: A Complement to Vector Calculus*, Academic Press, San Diego, CA, USA, 1997.

Weisstein, Eric W., *CRC concise encyclopedia of mathematics*, CRC Press, 2003.

Whelan, P.M., Hodgeson, M.J., *Essential Principles of Physics*, John Murray, 1978.

Whitney, Hassler, *Geometric Integration Theoy*, Princeton University Press, Princeton, 1957.

Index

E

F

G

H

I

L

M

N

O

P

Q

R

S

T

U

V

Z

❑❑❑